LOVE THIS BEAUTIFUL MUSICAL MATHEMATICAL UNIVERSE

LOVE THIS BEAUTIFUL MUSICAL MATHEMATICAL UNIVERSE

Lena Rabi Capapas, MD

ISBN-13: 9781546927297
ISBN-10: 1546927298

Library of Congress Control Number: 2017908764
CreateSpace Independent Publishing Platform,
North Charleston, South Carolina, USA

Dedication

This work is most lovingly dedicated to the memory of
Bishop Apolonio A. Capapas, Andrea Rabi Capapas,
and Domingo Gregorio Rabi Capapas.

Preface

A few years ago, I "shifted focus," or, in common terminology, I retired. I preferred the terminology of shifting focus because, according to the dictionary, to retire means to withdraw from action or danger, to stop working, to move away, to fall back, to recede or to retreat from battle, or to go to bed. None of these definitions applied to my intended new way of living. Although I had closed my medical practice, which I still miss, I certainly did not want to stop working, or retreat from life, or go to bed, though having the freedom to take a quick nap in the afternoon is unquestionably sweet.

"Shifting focus" seems a more appropriate phrase to me. It connotes a continuation of activity and of movement forward toward a fresh horizon, a new goal or two, a different field of interest or the realization of a dream long held and still unfinished. It flashes a mental picture of revving up the engine of creativity and of shifting gears to chase a compelling pursuit. It holds a promise of reviving a lust for life long buried in the busyness of carrying on a balancing act among career, family, and daily living. I do not mean to diminish the importance and depth of personal satisfaction that each of my juggled life devotions blessed me with. No, not at all. Still, this shift in focus has brought me a breadth of freedom that makes new activities engrossing, exciting, and enormously life-giving. In its own way, it is gently mellowing and kindly forgiving of the times when I had forgotten to stop and smell the roses along my old, harried way.

Consequently, I respectfully suggest that maybe, on some lazy Sunday afternoon, you might think about shifting focus as it applies to you. You will probably come up with more interesting thoughts than I have. Please try not to blame me if you feel a sense of urgency, if you can hardly wait to do it, and if all of a sudden, things as they are no longer satisfy.

It was with much trepidation, heaped with a bold measure of excitement, that this attempt at writing about this beautiful musical mathematical universe blossomed. I had studied and practiced medicine, so science had been my playground. But the area of quantum physics, a discipline of science that seeks to explain the nature and behavior of energy and matter at the atomic and subatomic levels, was darkly unfamiliar. Yet, a beguiling song lured me to its lair, and willingly, I stayed.

Wonderfully lost in its vast sea of knowledge, I felt a strong sense of responsibility to share what I had learned with my friends and family, using easily understandable terms. There took root a desire to have scientific information make reasonable sense where the experimental facts leave off and our daily lives begin. I realized that once we connect the dots to the best of our ability and become cognizant of the deep connection between the principles of physics—especially those of the incredibly successful quantum theory—and the art of living, we can transform our lives from ordinary to one of awesome awareness and magical joy.

In the last three years of my practice, I found myself voicing a feeling of being led down the road of "vibrations"—why, what, where, and how, I did not know. Although, by that time, vibrational medicine had already attracted several enthusiastic and dedicated practitioners, I had no direct knowledge of or practical exposure to the field.

One time, during a medical conference, Sharry Edwards, a pioneer of bioacoustics biology, identified the dominant vibration of the room and proceeded to hum the note out. Her voice seemed to grow louder and louder, soon filling the whole room as it resonated to her humming. The air seemed to expand. I felt as if I were being squeezed rhythmically and that the room was going to explode. The experience was, to say the least, unforgettable. It definitely blew my mind.

It would have been natural, then, to explore vibrational medicine. Instead, this shift in focus has taken me to the world of vibration of the ancients: to Hermes Trismegistus, the purported teacher of life principles embodied in the sacred texts of Hermeticism; to Pythagoras of Samos, the Greek philosopher, scientist, mathematician, and lyrist who supposedly learned Hermetic thought; to the Eastern philosophy of energy flow; and then, strangely, back again to the world of experimental facts, vigorously tested physics theories, quantum connectedness, and weird "action at a distance."

How incredibly bizarre! From medical science to ancient mysticism and occult knowledge, then back to the scientific hard facts of quantum physics and its never-ending enigma, my path took a jarring right-angle turn, became an indescribably mind-boggling twist, and finally ended in a surprisingly mystical landing on a harmonically entwined universe.

What a fantastic and wondrous ride it has been, and I invite you to join me through the pages of this book. Take it as nice and easy as you wish or as fast and furious as you are driven. The journey is all yours to create.

Acknowledgments

I wish to thank my husband, my sons and daughters, and my grandchildren for giving me the freedom to spend as much time as I needed on this book, which was as much a fun trip as a project. I also wish to thank my siblings and their families, my esteemed ancestors, and my extended family and its fully inclusive membership for their intricately knitted presence in my life. I feel blessed to have their love and inspiration.

I am also grateful to all my friends who have encouraged me; to my nephews and nieces for their artwork; to the giants of science, great and small, for sharing their gifts; to the ancients, whose wisdom has informed and guided me through the years; to NASA, whose multiple cosmological research arms provided the beautiful pictures of the universe gracing the pages of this book; to Professor Marc Vrakking and Dr. Aneta Stodolna, for their help toward my understanding of their work on the hydrogen atom; and to Sir Timothy Berners-Lee, for his brilliant and generous idea of sharing information with the world. To those who have helped whom I have not mentioned here, I thank you all from the bottom of my heart.

Introduction

In 1918, Max Planck, a German theoretical physicist, won the Nobel Prize in Physics for his quantum theory. Since the theory's inception, its predictions have been unfailingly proven in experiment after experiment, and its practical applications have been spectacularly successful. The quantum is awe-inspiring, to say the least. The simple light switch, computers, cellular phones, magnetic resonance imaging (MRI), lasers, electron microscopes, satellites, the Global Positioning System (GPS), the internet, and many other technological devices employ the principles of quantum mechanics.

In spite of the triumph of quantum theory, several curious observations in the physics laboratory, such as the "observer effect," a phenomenon in which the very act of observation affects the object being observed, and "quantum entanglement," where two or more entities that have once interacted are forever interconnected across space and time, have been found to be mysterious and difficult to explain and to believe. Furthermore, these phenomena demand the acceptance of consciousness as affecting the domain of the very small—so small as to be beyond the imaging capacity of the most powerful microscopes—the quantum energy field. And here is where the rarefied arena of physicists has split—big time.

A physicist I am indeed not. Upon embarking on this journey, aware of my ignorance on many levels about the subject matter, I had to constantly gather courage from the words of Socrates: "All I know is that I know nothing." In learning from the scientific inquiry of others and from wisdom handed down through the ages, I am emboldened to tie in the profoundly enigmatic quantum with the fascinating "gross" natural phenomena—the things, objects, and events that we see, hear, or feel—that are part of everyday life.

This exploration has only just begun, however, and the learning curve has been steep. Often I find myself drawing, or computing, or pacing back and forth to assimilate information that's just been crammed into my brain. There is so much to absorb that if overwhelmingly thrilling information were food, indigestion would already be the diagnosis given to this stomach-churning predicament: one that some of you might already be familiar with. Painless it is certainly not.

For this very reason, I have written this book in a manner that I hope will make it easier and smoother for those readers who might want to embark on a trip similar to mine and who are new to the world of vibrations and science, as well as those readers who are simply interested in dabbling in new ideas and modes of thought.

A few highly technical terms may be so matter-of-fact as to be second nature to certain people, while they may remain as unclear as the dense morning fog to others. Even though I may not define some of these terms immediately, I will provide clarification shortly after introducing the term. Readers may possibly skip a few passages. Still, some parts could contain bits of information on scientific discoveries and theories that other ideas build on, so a patient endeavor at comprehension is called for.

This does not have to be discouraging. I have made a conscious effort to repeat the same information a few times so that it will become more and more familiar. After some time, it all comes together, and you will find yourself no longer asking the question posed by impatient kids on a long trip: "Are we there yet?" Before you know it, you will already be there, having fun.

Throughout this book, I try to use the word "physical" to refer to that which follows the physical laws of nature, and I use "material" to refer to "gross" manifest matter—that is, all of nature that we see, hear, and feel. For the purposes of this work, I consider it important to distinguish between non-manifest physical and manifest material phenomena, although it would certainly be awkward to substitute the word "material" for "physical" in Olivia Newton John's song "Let's Get Physical."

You may also find statements in this work that may soon be considered mistakes in light of newly acquired information in the scientific world. And you may disagree with some of the philosophical views espoused in the book. For these instances, it would be immensely appreciated if you could deign to be kindly forgiving of my shortcomings and gently accommodating of the differences in the way we look at life.

You may side with me and learn with me, or you may totally reject this work as the confused, inconsequential ramblings of an old woman. I welcome your every response. Having lived to the golden warm September of my years, anything life throws at me now is proof that I am still engaged in the art and adventure of living; I take it with awe, gratitude, and mellow humor, if not with a grain of salt. Nevertheless, I most certainly hope that in your heart you will know that my intentions and efforts are truly pure and sincere.

I invite the skeptics and scientists among you, to whom anything that smacks of metaphysics belongs in the trash, to keep an open and curious mind and to continue reading, just in case something in the book happens to illuminate some shadowy and uncomfortable areas in your particular understanding of nature. I promise that this material, at the very least, contains recent advances in quantum biological science that are interesting and thought-provoking, even paradigm-changing—as they apply to our experienced reality.

According to the physicists Bruce Rosenblum and Fred Kuttner in their bestselling 2006 book *Quantum Enigma: Physics Encounters Consciousness*, "When experts disagree, you may choose your expert. Since the quantum enigma arises in the simplest quantum experiment, its essence can be fully comprehended with little technical background. Non-experts can therefore come to their own conclusions. We hope yours, like ours, are tentative." I have tried to be as tentative as I could, as much as I have taken pains to give expression to ideas that need to be recognized and given import, whether they be exhilaratingly fresh or so musty-old and forgotten that they startle as if new. Read on, hitch on to this ride, and let's have fun.

And remember: you are loved.

Contents

1
Cosmic Evolution

Cosmology is among the oldest subjects to captivate our species. And it is no wonder. We're story tellers, and what could be more grand than the story of creation?
—Brian Greene

Before the Italian philosopher, physicist, mathematician, engineer, and astronomer Galileo Galilei (1564–1642) produced an improved telescope that heralded the birth of modern cosmology, the earth was widely regarded as the center of the universe. The then-accepted Ptolemaic geocentric model of a stationary earth with the sun, moon, and stars revolving around it was in accordance with the cosmology of Aristotle, accepted as the truth and blessed by the Roman Catholic Church. This model of the universe was challenged when Nicholas Copernicus (1473–1543), a Polish scientist and Catholic cleric, offered a different view in his 1543 book *On the Revolution of the Heavenly Spheres.*

In the Copernican model, the sun became the center around which the earth and the moon revolved. This "heliocentric" stand was too revolutionary and too dangerous to support at that time. Those who advocated it were strongly criticized and persecuted by the Church. In 1600, the friar and polymath Giordano Bruno (b. 1548) was burned at the stake for asserting that other solar systems existed, with living beings in them besides.

In 1610, Galileo constructed a telescope after only hearing of its invention by Hans Lippershey, a German-Dutch master lens grinder and spectacles maker. Shortly after, Galileo found four moons orbiting Jupiter, which convinced him that the Copernican model was correct. He also observed craters and mountains on the earth's moon, sunspots, and heretofore-invisible tiny stars composing what then looked to be only a nebulous ribbon-like swath of milky luster across the sky, the Milky Way galaxy. This was corroborated by Christopher Clavius (1538–1612), Rome's Jesuit mathematician and astronomer (he finalized the modern Gregorian calendar originally proposed by Aloysius Lilius) and some of his fellow mathematicians at the Collegio Romano during Galileo's visit with him in 1611.

Galileo's public support of the non-geocentric model prompted a warning from a Catholic cardinal for him to refrain from defending and teaching the Copernican theory of heliocentrism after it was condemned by the Roman Inquisition in 1616. It was during this time that he wrote his seminal book *Dialogue Concerning the Two Chief World Systems: Ptolemaic and Copernican*, which earned him the famed title of the founder of modern physics. The work went against Catholic teachings and was banned shortly after publication in 1632.

He faced the Roman Catholic Inquisition's body of ten cardinals in 1633 and he was made to renounce his Copernican views. The Inquisition's proceedings ended in Galileo's house arrest in the comfort of an archbishop's palace in Siena, Italy, where he was free to gaze at the skies with his telescope. Later that year, he was permitted to continue his house arrest in his own villa in Florence, where he died in 1642.

After the trial, *Dialogue* was included in the Index of Forbidden Books and remained prohibited until 1835. Nonetheless, his ideas escaped the clutches of the Catholic Inquisition, since the book saw print in Latin in France in 1635. His postulations were published in Holland in 1637 and the volume was translated into English in 1661 as *Discourses and Mathematical Demonstrations Relating to Two New Sciences*.

BORN FROM A SINGULARITY

According to science, a "singularity" is an infinitesimally small point of infinite density and gravity where the known laws of physics break down and totally fail. The theory that the universe originated from such a singularity, which is strongly supported in modern cosmology, was introduced in 1927 by Monseigneur Georges Henri Lemaître (1894–1966), a priest, astronomer, and professor of physics at the Catholic University of Leuven (Louvain) in Belgium. He propounded the hypothesis of the "primeval atom," sparked by the equations of general relativity published in 1915 by the German-born mathematician and physics genius Albert Einstein (1879–1955), the solutions of which allowed for the appearance of singularities. (The general relativity theory was a necessary sequence to Einstein's theory of special relativity introduced in 1905; both theories are discussed at length in chapter 11.)

Lemaître theorized that the universe was birthed from this singularity in a sudden blast of hot and rapid expansion. The English astronomer Fred Hoyle mockingly referred to this as the "big bang" because the theory presented to the imagination a picture of an ear-splitting projectile explosion into space from a central tiny point, scattering matter all through space like shards and fragments flying from a violently shattered glass.

Cosmologists say it was not that kind of explosion. Rather, it was a sudden, very fast inflation of everything there was in the singularity. The size of the universe blew up by one hundred trillion trillion times in much, much less time than it takes to blink an eye. It continued expanding at a rate many times faster than the speed of light during a period termed the "inflationary epoch" and has been expanding ever since at a much slower rate. Regardless of its origin, one has to admit that the term "big bang" is dramatic and catchy, and that is probably why it stuck.

The idea of the universe starting from a singularity was not novel. In her 1999 book *The Secret Doctrine of the Kabbalah*, Leonora Leet (a professor of English and author) wrote that ancient kabbalistic cosmology, according to the respected sixteenth-century kabbalist Rabbi Isaac Ben Solomon Luria, teaches that the birth of the cosmos began exactly from this central point of dense singularity that suddenly expanded.

The kabbalah is a Hebraic mystical school of thought, the esoteric teachings and practices of which strive to define the relationship between an infinite, divine, mysterious, immutable, and eternal light and a finite material universe, especially human beings and our journey to return to the light.

During the Renaissance, the study and application of kabbalistic teachings in Christian belief became widespread, and Lemaître must have been very well versed in them, given his extensive range of scholarly studies for the Catholic priesthood which would have included physics, mathematics, and eventually, cosmology. His theory of a universe expanding from a primeval atom paints a mental image exactly like that of the Lurianic cosmic birth.

Even before presenting his concept of primeval atom cosmogenesis, Lemaître had already introduced his theory on the expansion of the universe. Written in French, his paper, "A Homogeneous Universe of Constant Mass and Growing Radius Accounting for the Radial Velocity of Extragalactic Nebulae," was published in a 1927 issue of *Annales de la société scientifique de Bruxelles*. Lemaître said that in an expanding universe, every galaxy would be receding from every other galaxy at a rate that would increase proportionately to their distance of separation. Thus, galaxies move away from one another ever faster the farther apart they are. He provided an estimated value of the rate of expansion of the cosmos, derived from calculations using data on distances and velocities of nebulae and galaxies collected by the US astronomers Edwin Powell Hubble (1889–1953) and colleagues.

ENTER THE ASTRONOMERS

Hubble, an American astronomer at Mount Wilson Observatory in Pasadena, California, announced at the January 1925 meeting of the American Astronomical Society that he had discovered that what had previously been thought to be nebulae were actually galaxies outside the Milky Way. It was an audacious claim. Up to that time, the Milky Way was generally considered to comprise the whole of the universe.

This information, scoffed at by some members of the scientific community then, excited a lot of scientists and helped propel the cause for astronomers. It encouraged the funding of research into extreme deep space and distant stars. The budget of the National Aeronautics and Space Administration (NASA) for 2016 was $19.3 billion, according to Forbes: over two hundred times the 1958 budget of $89 million dollars, equivalent to $742 million in 2014 constant dollars.

In October 2016, NASA estimated the number of galaxies in the observable universe to be two trillion: ten times more than previous estimates of 200 billion. The long-lived and beloved Hubble Space Telescope that was launched in 1990 has since sent back well over one million observations and numerous stunning photographs of the distant cosmos. In 2015, NASA released a photograph of GN-z11 just as it was, an infant galaxy 13.4 billion years ago—a mere 400 million years after the big bang.

THE FIRST BLUSH OF COSMIC DAWNING

The big bang left faint footprints in deep space. In 1964, a soft afterglow of background microwave radiation from primordial photons was serendipitously discovered by the American radio astronomers Arno Penzias and Robert Wilson of Bell Laboratories in New Jersey. Initially, they noticed a uniform noise on their radio, which they thought had originated from pigeon droppings which coated their equipment antenna. It turned out to be the ancient relic of the first light and sound of the universe.

The very early universe was fully made of radiant energy. There were only fast-moving electrons and quarks, the latter being subatomic elementary particles discovered not long ago. Soon, once the temperature cooled down to around 1,013 degrees Celsius (1,855 degrees Fahrenheit), the quarks coalesced to form particles: the positively charged protons, and the neutral neutrons. These subatomic particles constituted the intensely hot and inconceivably dense primordial cosmic soup through which light radiation could not penetrate.

At this point, atoms were not yet formed, since the attraction between protons, neutrons, and electrons was not quite strong enough to overcome the speed of their jostling. Further cooling to around 10,000 degrees Celsius (18,000 degrees Fahrenheit)—at about 300,000 years after the big bang—slowed the early subatomic particles down enough to allow them to fuse, forming clumps of particles with sparse spaces in between.

The variations in the temperature, particle speed, and density of matter in different regions were frozen in time, leaving a uniformly fluctuating background glow all over the heavens and beyond into the vast expanse of the void. This light radiation is what Penzias and Wilson picked up on their radio in the microwave spectrum of radio signals. Thus, the term "cosmic microwave background" (CMB) was born. They were awarded the 1978 Nobel Prize in Physics for this discovery.

Using sophisticated software, scientists from the University of California in Santa Barbara have converted the CMB "anisotropies" (i.e., non-uniformities) to sound via a technique known as sonification through an ingenious high-tech process, unmasking primordial sound-wave oscillations that are distinctly harmonically guided. Though the frequency of this sound is too low to be heard,

John G. Cramer, professor emeritus of physics at the University of Washington in Seattle, has scaled the sound 100 septillion times (10^{26}) upward and has posted it on the web.

In 1989, NASA launched the spacecraft COBE (the Cosmic Background Explorer) specifically to detect and measure the background microwave and infrared radiation and the variations in its density and temperature. Built to the tune of $160 million and involving a team of more than a thousand scientists and their helpers and several laboratories, COBE found exactly what it was looking for. Its massive mapped data reveal faint ripples in the temperature and density of the background radiation arranged in a specific pattern, satisfying the theoretical predictions of the big bang model and setting it on much firmer ground as the cosmic beginning. The disparity in temperature and density after the big bang could provide just the right physical conditions for the creation of stars, and thus of life itself.

The astrophysicists George Smoot, John Mather, and their colleagues took three long years to finely sort through the voluminous data from COBE. At the 1992 annual meeting of the American Physical Society, Smoot excitedly announced that English lacked the superlatives to convey what they had found. The team had observed fifteen-billion-year-old fossils they thought had been created at the birth of the universe. (In 2013, the European Space Agency's Planck mission estimated the universe to be 13.8 billion years old.)

Smoot's exuberance was justified. "Wholesky" maps (figures 1–2) reveal gorgeous pictures of the universe permeated by a uniform background with very faint fluctuations. The slight variations were boosted by image-enhancement techniques, which are now so advanced that the tiniest differences can be deciphered and clarified, and spectacular pictures are easily achievable.

As the astronomer and television (TV) host Carl Sagan (1934–96) wrote in his 1980 book *Cosmos*, "The Cosmos is all that is, ever was or ever will be. Our feeblest contemplations of the Cosmos stir us—there is a tingling in the spine, a catch in the voice, a faint sensation as if a distant memory, as if we were falling from a great height. We know we are approaching the greatest of mysteries."

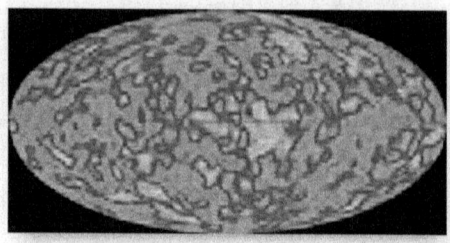

Fig. 1. The density fluctuations of the cosmic microwave background (CMB) shown with enhanced contrast between hotter, denser areas and cooler, rarified areas arranged according to harmonic intervals. Source: NASA Legacy Archive for Microwave Background Data Analysis (LAMBDA).

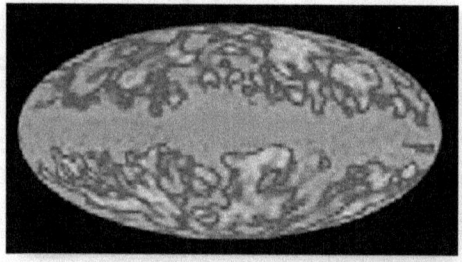

Fig. 2. The CMB sporting a belt created by light from the stars in the Milky Way's galactic center. Source: NASA LAMBDA.

The anisotropies of the radiation spectrum in various regions of space are thought to represent the contrast of matter's density and are light echoes of the primordial sound waves that bathed the baby universe. These minute variations are ordered according to harmonic intervals that result in musical overtones, the oscillations of which outline specific harmonic patterns and unveil a definitely symphonic principle guiding the creation of this beautiful universe. Mather and Smoot won the 2006 Nobel Prize in Physics "for their discovery of the blackbody form and anisotropy of the cosmic microwave background radiation." Their work cemented the foundation of the standard big bang model of cosmic evolution and started a new era of precision cosmology.

This thrilling new investigative sector of high-tech detection and imaging has presented to us stunning pictures of distant galaxies. It also has inadvertently exposed a hidden secret of the dark depths of space: in its bosom, the lovely, star-studded night sky is far from serene. In fact, it appears downright violent. A prime example is the fiercely intense radiation of quasar PKS 1127-145, which still shines with dazzling brilliance into every little celestial nook and cranny with a full spectrum of radiation, even across ten billion light-years.

Lemaître lived to witness the advent of state of the art high-precision cosmology and astronomy and the confirmation of his primeval atom hypothesis on the birth of the universe. He died at almost seventy-two years old, two years after Penzias and Wilson had their charmed encounter with antenna static from what they thought were pigeon droppings. Now, given the new information on the CMB from the Planck mission, the good monseigneur, wherever he may be, must be grinning from ear to ear.

Singularity, primeval sound, and primeval light all figure in the story of the birth of the cosmos according to both ancient accounts and to modern scientific discovery.

2

Heavenly Musings

Twinkle, twinkle little star, how I wonder what you are
Up above the world so high, like a diamond in the sky.
—Ann and Jane Taylor

Many people love singing this little ditty to Mozart's variation of an old French melody with little ones, and yet how much time do they really spend wondering? Probably less than inquisitive babies still full of fascination and curiosity about the world they are only beginning to get to know.

The secrets of the black moonless sky have long captivated professional stargazers and novices alike. The same mysteries must have cast a spell on you at some juncture in time, some way, somehow. Have you ever wondered what a star really is? What it is made of? Why it shines? Why it twinkles? How far it is from you? How it relates to you and to the rest of the planet? Why, in the first place, it is there? Many philosophers, astronomers, and starstruck watchers of the wild black yonder have pondered these questions before and reached conclusions that have added to the wealth of knowledge all successful stargazers have based their searches on.

In the very first few fractions of a second after the big bang, there were no suns, no galaxies, no planets, and no moons. Cooling of the cosmic soup calmed down the frenzied motion of the primary particles (consisting of electrons, protons, and neutrons) and allowed the formation of atoms: mostly the gases hydrogen and helium, and later, lithium. These early simple atoms subsequently stuck together to form small clumps that grew bigger into clouds of gas, then into collections of even more gigantic cloud formations. They grew so big that they started to collapse from their own gravity.

Soon, the densest central cores of the collapsed clouds became increasingly hot and more energetic as they became progressively condensed, fusing hydrogen atoms, which led to the release of more energy, which then caused the stellar gaseous content to expand. This expansion counteracted the gravitational pull exerted by the dense cores and stabilized the newly formed stars. Many of them became the earliest massive stars of the universe, up to one thousand times bigger than our sun. The first light that graced the heavens came from these first-generation stars.

In the gargantuan gaseous mountains, larger stars gobbled up smaller stars, adding to their masses in the process. This resulted in an increase in their gravitational force, causing a contraction that drew in smaller stars and brought everything into a spinning harmonic movement. This nascent galaxy slowly assumed the shape of a disk, with the strongest gravitational force in its central core, which was surrounded by several spiral arms of harmonically bound stars and likewise entwined planets orbiting around them.

The Milky Way is one of several hundred billion galaxies accounted for by technology, though the number could easily be infinite, according to the equations of quantum mechanics. The number of stars in the Milky Way, an average-size galaxy, is estimated to be around four hundred billion, although estimates vary wildly. As noted earlier, NASA estimates the observable universe to have some two trillion galaxies, and so, if we multiply four hundred billion stars in an average-size galaxy by two trillion, the mind totally shuts down in self-preservation. (Seriously, the total number would be around 800 sextillion, or 800 times 10^{21}.)

We see only around five thousand stars with the naked eye at any one time, because we do not see the sky on the other side of the earth. We might see more if we traveled to the far northern or far southern hemispheres to count the stars on moonless nights. The reason is because more people live in the mid-northern and mid-southern latitudes, which means more city lights and more particle air pollution, which compromise our visibility. We can count even more stars at the poles, where stars never rise or set and all stars visible in that part of the sky stay up day in and day out. The odds are even better at the South Pole, where the elevation of 2,835 meters (over 9,000 feet) above sea level allows more of the sky to be visible above the horizon.

So, if we traveled from the North Pole to the South Pole to count all the stars during clear, inky-black nights, we would end up with a total of around ten thousand stars visible to the naked eye. According to the famous English theoretical physicist and cosmologist Stephen Hawking (1942–2018), if we imagine that a star is a grain of salt, all five thousand stars visible from on one hemisphere of the planet would fit in a teaspoon. In contrast, all the stars observed by sophisticated gadgets would look like a giant ball of salt eight miles (12.8 kilometers) wide if rolled into one. By now, you must be starting to feel infinitesimally small and inconsequential. There is hope, though: you and I are stars in our own right. We are children of those points of light twinkling in the dark. We are stardust.

A star like our sun, being composed of gas, has a hot furnace for a center, the product of enormous amounts of heat and light radiation during nuclear hydrogen fusion reactions that convert hydrogen into helium. This is the reason why a star shines so brightly from far across the sky. As the tiny point of light travels through turbulent atmospheres on its way to the earth, the light gets bent as it gets refracted and reflected by atmospheric particles and is sent zigzagging here and there. During some of those "zigs" and "zags," the light strays outside of the field of vision, giving the illusion that the starlight is intermittently disappearing—thus flickering and twinkling.

The more atmosphere the starlight has to go through, the more wildly it glitters and the more it gets broken up into varied colors and hues. Jewelers utilize this principle with their full-spectrum lights, which get refracted and reflected by an expertly cut diamond solitaire, showing off the dazzling dancing sparkles coveted by gem lovers.

As a star grows older and much of the hydrogen fuel is exhausted, the nuclear fusion slows down to a point where the star's energy production can no longer support the gas expansion necessary to counteract the pull of gravity exerted by the core's mass. It then starts to collapse onto itself, getting smaller and denser. But further compression renews nuclear fusion, turning helium into carbon and releasing heat and light, so, the star expands again, reaching up to 100 to 1,000 times the size of our sun: between 100 million to 1 billion kilometers in diameter. Because of the star's massive expansion, its outer gas layers are cooler, and the star radiates light in the red color spectrum. It is now a red giant.

As the core helium is exhausted, the star begins to contract once more. In the course of time, the outer and cooler layers of gas dig down into the hot core in a convection-driven flow, disrupting the core and pulling some of its contents up to the outer layer. The core sometimes bursts with such energy that portions of the dying star's outer layers are blown off, creating a shroud of gas and core dust around it. The gas and dust drift away to form nebulae, the birthplaces of new stars and galaxies, leaving the stellar core bare.

A star that is less than four times our sun's mass does not generate enough energy to start carbon fusion, so helium remains as its core. As the outer hydrogen shell and the outermost gas layers expand and are cast off into space, the hot helium heart of the star becomes exposed. The star is now a full-fledged, fuel-exhausted white dwarf. A white dwarf is what our sun is destined to be: a mere earth-size remnant of what is currently a giant ball of fire. (The earth has an equatorial diameter of 12,756 kilometers or 7,926 miles.) Even though the star will still be fiercely hot at this stage, its luminosity will be faint because of its small size. It will continue to burn for as long as its helium supply holds, beyond which it will gradually cool down and slowly fade away unceremoniously, hypothetically turning into a black dwarf.

A white dwarf belonging to multiple star systems, such as a binary system (for example, Sirius A and B) or a triple system (16 Cygni), follows a different path. The largest white dwarf in the star complex attracts the outer material (mainly hydrogen) of its neighbors and starts to engulf it, gaining more fuel. With time, the white dwarf's outer layers gather enough hydrogen to ignite a sudden fusion reaction, and it explodes into a nova.

A collapsing massive star over eight times the mass of our sun charts a different course. As it caves in on itself, its increasing density promotes the resumption of escalating nuclear fusion. The more massive a star, the denser and hotter its core and the greater the complexity of its nuclear fusion. The complicated reactions result in heavier elements, mostly iron; this process requires exceedingly more heat and energy to fuse and quickly exhausts the star's energy supply.

Fusing iron does not yield as much energy as fusing hydrogen, and the delicate balance between pressure-induced expansion and mass-driven gravitational shrinking is upset. Suddenly, the heart of the star implodes, and the star collapses in the twinkling of an eye. The hot, mostly iron core becomes so tightly squeezed that the repulsive forces of the iron's atomic electron shells compel an abrupt energetic rebound. The dying star suddenly explodes and turns into an intensely bright supernova that could outshine an entire galaxy. A supernova differs from a nova in its extreme brightness and much larger size. Most importantly, it differs in the content of the extruded material. The inner core material as well as the outer gas layers are blasted away in the explosion; in a nova, only the outer layers are blown off. A supernova widely scatters gas and heavy elements all over space: the source of all natural elements found on earth and in the human body, including the iron in our blood and the calcium in our bones.

Precious metals and stones like gold and diamonds were formed eons ago in this extreme heat and pressure not duplicable on earth. Consequently, they are exhaustible and thus expensive. Although scientists may discover anywhere between twenty-five to fifty supernovae each year, only a single supernova explosion occurs every fifty years in the Milky Way on average, most of them too distant to be seen. The remnant of the last visible explosion in 1680 is shown in figure 3.

If the core of a supernova is 1.4 to 3 times the mass of our sun, the star continues to collapse to such a high energetic stage where protons combine to form neutrons, and it becomes a neutron star. This is a phenomenally dense star that has such gravitational force that it is able to cannibalize any companion star near it. A neutron star's immense energy generates a powerful magnetic field that emits a beam of radiation from each magnetic pole that sweeps in and out of the line of vision in regular pulses. It does so because as the star rotates around its own axis, which is not in alignment with the magnetic pole axis, the paths of the light beams rotate with it. Astronomers call this kind of star a pulsar.

If the inner stellar core is bigger than three times the mass of our sun, the supernova may develop an iron core so great that no more fuel for further nuclear fusion is available. In other words, the supernova is completely burned out. No more pressure is generated in the core for outward expansion, and consequently, the star's own gravity forces it to collapse completely upon itself, leaving a remnant so infinitesimally small as to have zero size, all the while retaining a definite mass that is equivalent to the mass of the star that died—but now of infinite density.

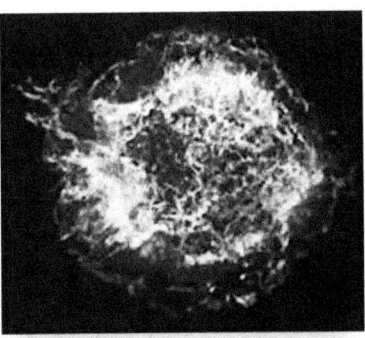

Fig. 3. The beautiful remnant of supernova Cassiopeia A, which exploded in 1680, as captured by NASA's Chandra X-Ray Observatory. Source: NASA / Chandra X-Ray Center (CXC) / Smithsonian Astrophysical Observatory (SAO).

It has mass, but it has no size; it has no size, but it has infinite density. It has now become a singularity—a celestial sinkhole from the gravitational pull of which nothing can escape, not even light. We call this a black hole, a term coined by John Archibald Wheeler (1911–2008), a prominent theoretical physicist at Princeton University who collaborated with Einstein in search of the grand unified theory of everything: the merging of the four fundamental forces of nature: gravity, electromagnetic, and nuclear forces, both strong and weak.

Karl Schwarzschild (1873–1916), a German astronomer and physicist, worked on solutions to the general relativity equations as soon as Einstein introduced the theory in 1915. His solutions described a center singularity and a critical gravitational radius from this singularity—the Schwarzschild radius, popularly known as the event horizon. Sagittarius A, the Milky Way's black hole, has a mass 2.5 million times that of our sun and an event horizon many millions of miles in diameter. The threshold of this event horizon marks the point of no return for everything that enters it at or below the speed of light. Since not even light (or sound) is capable of escaping the gravitational pull of the singularity and its event horizon, scientists and their equipment have never seen or recorded them, since they utilize the frequencies of light or sound for their observations.

The presence of a black hole singularity is not directly observed but is deduced from its gravitational effects on galaxies and from the light radiation from the contents of its "accretion disk," the immediate region furiously swirling around the black hole's event horizon. In this region, massive numbers of colliding galaxies are in preliminary stages of being sucked into the whirlpool of the event horizon with stupendous speed, generating so much energy that the radiation makes this mammoth black hole "meal" shine. Twisting magnetic fields inside this accretion zone sometimes release a bright jet of light which is seen as coming from the center of the galaxy. This light jet is called a quasar.

3
The Birth of Quantum Physics

A new scientific truth does not triumph by convincing its opponents and
making them see the light, but rather because its opponents eventually
die, and a new generation grows up that is familiar with it.
—Max Planck

This prescient statement could have been its speaker's epitaph had
he not enjoyed a long and productive life, having lived to the ripe
age of eight-nine: certainly, long enough to relish the triumph of
his discovery of the quanta of energy. The newly introduced
quantum theory clashed with the reality backed by classical physics
so boldly that it was fiercely debated, anguished over, and firmly
rejected by many physicists for a very long time, never more
passionately than by the theory's very own father, Max Karl Ernst
Ludwig Planck (1858–1947). (Author Manjit Kumar's 2008 book
Quantum: Einstein, Bohr, and the Great Debate about the Nature of
Reality offers a great account of the budding quantum theory.)

In the very early twentieth century, Max Planck, renowned in
Germany as a theoretical physicist, was studying a problem of light
emission from a "blackbody," which confounded physicists. A
blackbody is an object that is ideally a perfect absorber and emitter
of light radiation, originally conceived and aptly named by another
well-respected German physicist (and discoverer of the elements
rubidium and cesium), Gustav Robert Kirchhoff (1824–1887). In
actuality, however, there is no perfect absorber. The cosmic black
hole and the familiar graphite (a form of carbon not dense enough
to become a diamond) come close to being near-perfect.

For non-physicists, a less-than-perfect blackbody sufficient for catching on to the phenomenon is a lump of coal. At room temperature, the coal emits radiation in the infrared spectrum. Infrared light is not visible to the naked eye, so the coal looks black. Still, we can feel this subtle, low frequency energy. As an example, if you put your palm next to your cheek without touching, you can feel the warmth of the infrared radiation from your hand. Yet, you do not see your hand glowing. If you are a lady who suffers from the "cold hands, cold feet" malady, you will lose out on this experience. You might have to borrow somebody else's hand, preferably a guy's, since by nature, men have warmer hands than women.

During that time in Planck's career, theoretical calculations utilizing classical physics—the laws of which rule the macroscopic, everyday world—predicted that as the temperature increased and the wavelength of light got shorter and the frequency of the waves of light became higher, the amount of energy given off would become proportionately more intense, ultimately reaching infinity beyond the ultraviolet spectrum. Following this logic, a lump of coal would get hotter and hotter until bursting into an intense ball of ultraviolet radiation, instantaneously losing its heat. That would simply be a nonsensical outcome, an "ultraviolet catastrophe," as scientists called it. This was the very problem besetting Planck.

The ultraviolet catastrophe went against what we experience in reality. A burning object does not explode into a ball of ultraviolet fire and die right there and then. Mother Nature appears to handle radiation in a gentler way. The peak intensity or brightness of light radiation released at extremely high frequencies and at fiercely hot temperatures always drops upon reaching very high frequencies. That is why a "red-hot" coal in the lower frequency spectrum feels hotter and more intense than a "white-hot" coal in the higher frequency spectrum closer to the ultraviolet, the total intensity of radiation having dropped from its peak. Heat from our sun is only 7–8 percent in the ultraviolet spectrum and 49 percent in the invisible low-frequency infrared spectrum. The remaining energy is in the visible low frequencies and long wavelengths.

In his blackbody "thought experiment,"[1] Planck imagined an ideal, well-insulated hollow black container with a hole in its wall big enough to let light in and small enough to limit the escape of the radiation to some extent so that light could be reflected back and forth in the cavity until it was completely absorbed by the black walls. Planck knew that if the theoretical blackbody was heated, the radiation emitted by the black walls would achieve equilibrium in the cavity, and whatever radiation escaped through the hole would be representative of all the wavelengths or frequencies in equilibrium inside the cavity.

When a blackbody is heated, the radiation it emits varies in frequency, depending on the temperature reached. That variation ranges from the infrared through all the increasing visible frequencies until the frequency reaches the ultraviolet range, which the eye can no longer see. The hotter it gets, the more intense the radiation it emits at all frequencies and the brighter the light you see—up to a point. We watch the coal glowing from cherry-red to red, then to orange, and then all the way to dramatically intense and blindingly brilliant bluish-white as the wavelengths get shorter and the frequencies get higher. That's before light finally disappears from human sight as the invisible ultraviolet radiation.

The ultraviolet catastrophe was not the only problem bugging the scientific community. Wilhelm Wien (1864–1928), a physicist at Berlin's Imperial Institute of Physics and Technology (Physikalisch-Technische Reichsanstalt, or PTR) optics laboratory,

---

[1] A thought experiment is an investigation into the nature of things performed in the imagination in which ideas, hypotheses, and theories are laid out as stories or situations, taking into account all their ramifications and consequences.

found that the maximum intensity of a specific energy radiation peaked at a specific wavelength, known as the peak wavelength. As the heat of the blackbody was increased, the radiation's peak wavelength got shifted to even shorter wavelengths or higher frequencies (known then as Wien's law of displacement). But further experiments by his colleagues using better blackbodies revealed that at the much longer wavelengths of the infrared spectrum, Wien's law overestimated the intensity of radiation. They found that the intensity of light was proportional only to temperature, and the wavelength no longer mattered. Wien's law of displacement failed at longer wavelengths.

The physics community had a big problem. A basic theory had to be found to explain the ultraviolet catastrophe and the failure of Wien's law of displacement. Max Planck, a reserved man by nature and upbringing, expressed his desperation by stating that he was ready to sacrifice "every one of my previous convictions about physical laws" in order to find a theoretical interpretation of the observations.

By that time, the Scottish mathematician James Clerk Maxwell (1831–79) had already discovered electromagnetism. Armed with knowledge from electromagnetism that an electrical charge can absorb and emit only the radiation that oscillates at the same frequency as its own, and faced with the obvious breakdown of Wien's law of displacement at very long wavelengths of infrared radiation, Planck went to work in a do-or-die frenzy, throwing his precious reputation and long-held views to the wind.

Max Planck constructed an imaginary blackbody with a wall made of vibrating molecules—essentially harmonic oscillators, one for every frequency of oscillation—that started to vibrate when the blackbody was heated. The energy released from each oscillator activated by heat would be released into the cavity and would be reabsorbed by a same-frequency oscillator. At a constant temperature, the emission and reabsorption of energy by the harmonic oscillators would achieve equilibrium, and exactly the same amount of energy would be absorbed as was emitted.

It was at this point in his scientific career that Max Planck was compelled to accept the reality of the existence of atoms, which Ludwig Boltzmann (1844–1906), a Viennese pianist who became an influential physics experimentalist, theorist, and teacher, held to be responsible for dictating the properties and behavior of gases in microscopic events. Planck had rejected the idea for a long time because of the absence of indisputable proof.

The physics community had already accepted the fact that gas was composed of atoms that behaved according to the second law of thermodynamics when inside a container. This law states that a system always evolves from a state of maximum order (lowest entropy) to maximum disorder and randomness (highest entropy). At maximum entropy, the atoms composing a gas are at their greatest state of randomness and disorder.

Everything in nature goes through the same undoing over the course of time—from order or low entropy to disorder or high entropy. Ever so sadly, a once-pretty face wrinkles and sags, and a well-developed athletic body gets arthritic and deformed through the years. And the eggs that you just dropped will remain broken, with their yolks splattered all over the floor; and will not regroup to become whole eggs again. You will just have to go back to the store if you want boiled eggs. Some effects of entropy are a little kinder. The snow melts from your driveway in the morning sun, so you don't have to shovel it. The scent that my friend sprays on her skin dissipates into the atmosphere, saving me from unleashing an embarrassingly wet and hearty sneeze.

Entropy was at work in Planck's thought experiment. As he turned up the heat of his blackbody, the vibrations increased and the random motion of the atoms in the chamber intensified. The amount of radiation released from and absorbed by the blackbody wall in an exchange with the light radiation from the chamber also increased. At a constant temperature level, a state of equilibrium between energy emission and absorption would be reached. This "steady state" was the same state of maximum entropy of atoms in Boltzmann's container. Planck had uncovered a hidden law.

Applying this law to his calculations, Planck discovered that he could arrive at a formula for the distribution of radiation in his blackbody if the energy absorbed and emitted by his oscillators was of the same frequency as the oscillators themselves. His calculations led to the conclusion that he could derive a formula only if he considered the energy at each light frequency to be composed of equal and indivisible individual units. He called these units "quanta," from the Latin word quantus, meaning "how much." He also found that he could derive a measurement of the energy E at each frequency v (sometimes expressed as f) by multiplying that particular frequency with a quantum of action constant, known as the Planck constant[2] (also Planck's constant), a minute number (6.626 divided by one thousand trillion trillion, or $6.62607004 \times 10^{-34}$ joule-seconds), which he arrived at in frustration, a "fudge factor" commonly applied by physicists to make their calculations fit their observations. This quantum of action constant is represented by h in his equation. It is a

---

[2] The Planck constant is one of five constants of nature—the others being the speed of light, Boltzmann's constant, Coulomb's constant, and the gravitational constant—from which Planck developed units, now known as the Planck scale, by "normalizing" their numerical values to 1. He did this for the purpose of having a universal system to use in physics equations that would, as he said, "retain their meaning for all times and for all civilizations, even extraterrestrial and non-human ones, and can therefore be designated as 'natural units.'" Expressed as Planck length (10^{-33} cm), time (5.39×10^{-44} s), mass, charge, and temperature, they can describe you, me, and anything else in the universe in an equation. The Planck length is so tiny that an invisible proton is 100 million trillion times larger than a Planck length. Planck time is the time it takes a photon, traveling at the speed of light, to traverse a distance of one Planck length. No scale of measure can be smaller than the Planck scale, as beyond it, classical physics (which describes life as we know it) theoretically fails and quantum mechanics takes over.

proportionality constant intended to limit the energies of his oscillators and their radiation to multiples of hv (Planck constant x light frequency). Thus, Planck's "E = hv" got its start to fame as the very first equation of quantum physics.

Using this equation on the ultraviolet catastrophe problem, Planck found that the blackbody radiation peaked at a certain frequency. The hotter the blackbody got, the higher the peak frequencies became. Then, at exceptionally high frequencies and short wavelengths close to zero, the radiated energy density also approximated zero because the energy obtained according to the equation was too big to be emitted. There would not be sufficient energy to excite the oscillators. There would be no light radiation at the ultraviolet level.

Quantized energy radiation means that there is a specific energy for every specific frequency emitted. This limitation of energy per specific frequency means that the radiation intensity or brightness of light can only be raised by increasing the quanta or number of energy units (later identified as photons) emitted.

One quantum of light with extremely high frequency and extremely short wavelength has extremely high energy. Even so, increasing the frequency does not necessarily translate to greater total intensity or brightness of light. As the heat in the blackbody rises and pushes light toward the ultraviolet, the frequencies and the frequency-dependent energy increase. In contrast, the total numbers of detectable specific-frequency emissions (or quanta of energy) decrease. Therefore, towards the ultraviolet level, fewer high-frequency oscillations and thus fewer quanta of light energy are detected compared to the total quanta of lower frequency radiation in the rest of the light-energy spectrum. Consequently, the total brightness of light radiation drops off.

The shutdown of the emission of quantized energy at very high frequencies effectively prevents the continuously increasing radiation to infinity that classical physics predicted. The anguish over the ultraviolet catastrophe was finally allayed. Applying the quantum of action constant h to Wien's law of displacement also solved the problem of the law's failure at very long wavelengths of the infrared light spectrum.

Physicists should have cheered. They didn't. At the December 1900 meeting of the German Physical Society in Berlin, Planck's new theory, which chopped up light energy into indivisible quanta with his constant h so that his blackbody oscillators could absorb them, got a less-than-enthusiastic response from the community. Many colleagues thought his theory inconsequential, just a fudge factor, a clever mathematical maneuver to arrive at the correct answer. The theory was further undermined by the attitude of Planck himself, who, though overjoyed and relieved, paid no more attention to his quanta after it satisfied his quest for a theoretical formula to solve the scientific dilemma of his time.

Little did Planck realize the significance of his discovery. Never did he suspect the compelling and monumental revolution in physics and the seriously big-time impact on the human race his quantization of light energy would create. Luckily for humanity, Planck's introduction of the quantum was not lost on Einstein, who was famous among his contemporaries for his strong intuition and his dedicated passion for delving far beneath the obvious into the nature of reality. He often derived insights from the utterly simple, even hidden, truths in physics as well as in his immediate natural surroundings.

While Planck was introducing his oscillator-dependent quanta of energy to the physics community, Einstein was working at the patent office in Bern, Switzerland, as a humble technical expert, third class. His environment was a world apart from the high perch of his scientific counterparts in prestigious universities. Nonetheless, his fairly light workload allowed him time to devote to his own interests, primarily physics. In addition, the discipline imposed by his empathetic if not doting superior, Friedrich Haller, in the critical review of patent applications finely honed his natural scrutinizing ability.

While Einstein was reviewing Planck's blackbody experiments, he became intrigued by the way Planck's calculations on radiation paralleled his own on atomic motion in liquid, the subject of the doctoral thesis he was working on. Ever inquisitive, he questioned whether the emission and absorption of Planck's light-energy quanta depended solely on the frequency of the oscillators in the blackbody and not on the nature of light itself. Planck had not bothered to think of all possibilities.

Quantizing the oscillators worked for Planck's calculations once he'd found h, and he was content with that, so quantizing light itself did not cross his mind. After all, he had just accepted the radical concept of atoms, forced by the necessity of arriving at a formula to explain his blackbody-radiation theories. On top of that, his orderly universe of classical physics established by Sir Isaac Newton (1642–1727), the great English mathematician, physicist, and astronomer to whom we owe the laws of gravity and motion that describe our day-to-day experience, had just been sorely violated and turned upside down. Planck was quite ready for a much-needed rest.

It was a different story with Einstein, who was experiencing exactly the opposite reaction. He was galvanized by Planck's light-energy quanta to perform intensive thought experiments of his own. He mentally constructed a blackbody with walls consisting of electrons and atoms, rather than the oscillators that Planck used. He then filled the chamber with gas molecules (composed of atoms) and freely moving electrons.

Heating the blackbody would make the atoms and their electrons in the walls and the chamber vibrate further, thus increasing the random motion and disorder of the blackbody's components. A steady-state equilibrium at a specific temperature would then be reached. At this point, maximum entropy of the atoms and their electrons would occur, which would be constantly moving here, there, and everywhere in the course of emission and absorption by the blackbody. Below is an excerpt from Einstein's translated paper on light, titled "A Certain Problem Concerning the Theory of 'Black Body Radiation'":

> We begin by applying Maxwell's theory of light and electrons to the following situation. Let there be a cavity with perfectly reflecting walls, filled with a number of freely moving electrons and gas molecules that interact via conservative forces whenever they come close, i.e., that collide with each other just as gas molecules in the kinetic theory of gases.

The electrons and atoms in his theoretical blackbody behaved like the atoms of contained gas in maximum entropy, with the energy inside the system acting as if it were fully composed of quantized particles. In this condition, his blackbody would already contain quantized energy, which then, would no longer need to be chopped up by Planck's blackbody oscillators.

Then, a brilliant thought hit his superconvoluted and super-connected brain: maybe light itself was quantized, and therefore, electromagnetic energy would be radiated in discrete quantized packets. Planck's light-energy quanta exist as such from the very start, and not after having been cut up to fit the frequencies of the blackbody oscillators. The electrons in the cavity and in the walls of the blackbody would emit and absorb light as discrete individual units. The ensuing formula, although originally written differently by Einstein, essentially contained the same information as Planck's $E = h\nu$. It was exactly Max Planck's successful quantum formula, notwithstanding Planck's own unawareness of the quantization of light itself.

Einstein's light quantum became known as the "photon" in a 1926 letter to the journal *Nature* by Gilbert Lewis, an American physical chemist. On the heels of Planck's light-energy emission and absorption quanta, Einstein's shining photons lighted the way for the budding discipline of quantum physics. In 1905, Einstein further clarified Planck's quanta through his thought experiments on the "photoelectric effect," the expulsion of electrons from metal by a beam of light of a certain required frequency, called "threshold frequency," measured in hertz. When an electron absorbs energy from a specific-frequency photon, the energy boost gets the electron excited. The electron's frequency becomes higher than and different from the threshold frequency required to keep it bound to the metal. The electron is therefore ejected.

The photoelectric effect is solely dependent on the frequency of the light beam and not on the intensity or brightness of the light. This could only mean that the frequency of the photon is proportional to the energy of the electron particle that is ejected, an interpretation that is compatible with an electron particle–like character of light. The ramifications of this discovery have yet to see an end. From the various technological inventions that use the photoelectric concept (such as solar panels and night-vision equipment) to our increasing understanding of how life works on earth, the concept seems to reveal more than Einstein probably ever predicted at that time.

For example, when sunlight photons strike a solar cell (which is usually made of silicon) in a solar panel, they knock electrons loose from the silicon atoms. The freed electrons flow through a conductor attached to the positive and negative sides of the cell, generating an electric current. With night-vision technology, the photoelectric effect allows light from a single photon to be multiplied several times by the release of more electrons, which would then emit more photons (known as photo-amplification and photomultiplication), thus increasing the brightness of light and enhancing our ability to see in the dark.

Photoelectric effect fundamentally figures in the performance of photosynthesis (discussed later), during which plants boast a level of superconductivity not achieved by high technology. Photoelectric effect is responsible for low-level light emissions in living systems, called biophotons, that are evidently involved in vibrational signaling between cells, tissues, organs, organisms, and within groups of organisms in order to implement their different functions—from admiring a beloved face to the synchronous dance of dolphins among the foamy waves of the deep-blue sea. Biophotons are also being studied for the possible use of light vibrational frequencies in the detection and treatment of diseases.

This achievement earned Einstein the Nobel Prize of 1921 "for his services to Theoretical Physics, and especially for his discovery of the law of the photoelectric effect." It was the only Nobel Prize in his long and illustrious career.

Interestingly enough, Planck strongly rejected the idea of light existing as isolated quanta and resisted the quantum movement for a long time. The revolution he fathered, however, just as strongly resisted dying: it spread like wildfire and blossomed in the thought experiments of his colleagues as well as in the laboratories of academia and industry. Watching his 1918 Nobel Prize–winning quantum of energy theory flourish in the scientific community and witnessing various inventions based on the theory proliferate in the commercial industrial sector, Max Planck finally learned to live with his legacy, albeit reluctantly.

4
The Atom

Matter is mostly ghostly empty space.
—Sir Arthur Eddington

Democritus referred to it with the Greek word *atomos*, signifying indivisible. He had this idea that cutting a grain of sand in half again and again must reach a point where the grain can no longer be divided, and one would end up with an indivisible and invisible unit of matter: an atom.

Born around 470 BC, Democritus was one of the greatest thinkers humanity has ever known. In his philosophical point of view, all things around him had to consist of basic and fundamental building blocks, though with a remarkable difference. The atom, according to him, was imbued with "specific properties" that enabled it to behave and interact with other atoms in certain ways to smoothly form and beautifully shape reality. That was a long time before any experimental indication of the chemical and quantum binding properties of the atom. He was way ahead of his time, even modern by today's standards.

In fact, Democritus's ideas paved the way for both ancient and modern alchemy. The original alchemists, the trailblazers of present-day chemistry, tried and failed to transmute lead and other base metals into gold, but not for naught. Even though they failed to have gold gleaming in their hands, the golden nuggets of knowledge they gained never lost their luster in the laboratories of chemists and physicists, today's quantum alchemists. The latter's mastery of the transformation of elements during atomic nuclear decay with release of radioactive particles, such as alpha and beta particles, stemmed from those experiments of long ago.

CHOPPING UP THE INDIVISIBLE ATOM

A tortuous and colorful history surrounding the term "electron" preceded the discovery that the atom is not solid; that it has parts. Around 1000 BC, the ancient Greeks had an active trade by sea with the Baltic region. One of their imports was the semiprecious stone *elektron*, popularly known today as amber, the fossilized resinous sap from a certain kind of pine tree, abundant deposits of which are found along the Baltic shore. They noticed that rubbing amber against fur resulted in an attraction between them. They also observed that some light objects such as hair and feathers clung to the rubbed amber. This curious characteristic was initially thought to be unique to the gemstone. Since then, this property has been found to be present in many other natural and synthetic materials as well.

Undisturbed objects in nature have neutral charge, the atoms composing them having a net zero charge. But different materials possess different degrees of binding strength or affinity to their very own electrons, and their readiness to give up their electrons depends on this property. When two objects are rubbed against each other, the friction created frees the relatively looser (bound) electrons of one object and transfers them to the other object with stronger electron affinity, resulting in charge imbalance.

Familiar objects with loose electron binding, starting from the loosest, include very dry hands, leather, rabbit fur, glass, human hair, nylon, wool, fur, lead, silk, aluminum, cotton, steel, wood, and amber. The tumbling of clothes in an electric or gas-heated clothes dryer generates friction and produces static cling in nylon clothing but not in cotton because the latter has a stronger hold on its electrons than nylon has.

In 1600, the English scientist William Gilbert, a physician to Queen Elizabeth I, wrote about this peculiar attraction of rubbed objects. He coined a new Latin word for the phenomenon, the adjective *electricus*, based on the Latin word *electrum*, a pale gold alloy of gold and silver, which, in turn, was a derivation from the Greek word *elektron* or amber, which is also gold in color. *Electricus* literally translates to "resembling amber." In the mid-seventeenth century, the word became anglicized into "electric," again an adjective that Sir Francis Bacon (1561–1626), an English statesman, philosopher, scientist, and author, used for materials that displayed the attractive property of rubbed amber. Sir Thomas Browne (1605–1682), an English polymath, physician, and author, used the noun "electricity" to describe the specific quality that would allow substances to attract light material: a connotation quite different from today's noun, which speaks more to the phenomenon, its cause, and its mechanism of action.

In 1874, George Johnstone Stoney, a famous physicist at the Queen's University of Ireland, contended that electricity came in a discrete negatively charged unit, which he later dubbed an "electron," a combination of the words "electric" and "ion." The word has now gone back phonetically closer to the original term, elektron. Although still related, the new term has now acquired an entirely different meaning.

Stoney's idea paved the way for Sir Joseph John Thomson's discovery of the electron. (J. J.) Thomson (1856–1940) was an influential physics Nobel laureate and Cavendish Professor of Physics at Cambridge University in England. In 1904, Thomson suggested that the atom was a massless sphere of positive charge in which negatively charged electrons were incorporated in a concentric formation, in the same way that plums are embedded inside an English pudding. In this model, the electrons accounted for all of the atomic mass. This meant that the atomic weights of different elements had to be limited by the number of electrons that each of them possessed.

A problem arose when scientists realized that there were far too few electrons to account for the calculated weight of an element; thus much of that weight had to come from the rest of the atomic sphere. This negated Thomson's description that the atom was massless except for the electron mass. Furthermore, his plum-in-a-pudding atom faced another challenge posed by the copious experimental output of Ernest Rutherford, First Baron Rutherford of Nelson (1871–1937), a young, highly energetic, and personable British–New Zealander who was a previous student of his at the Cambridge University Cavendish Laboratory.

THE PHYSICS OF THE ATOMIC NUCLEUS

A chemist, physicist, and premier experimentalist, Rutherford was awarded the 1908 Nobel Prize in Chemistry "for his investigations into the disintegration of the elements, and the chemistry of radioactive substances." He found that radioactivity was a phenomenon of transmutation of one element into another during atomic decay, resulting in the emission of alpha and beta particles. This work, done in collaboration with his colleagues, birthed nine important papers on radioactivity and transmutation within only two years.

In 1907, after Rutherford's nine-year stint as a Macdonald Professor of Experimental Physics at McGill University, in Montreal, he settled again in England at the Victoria University of Manchester. There, he continued his work on the scattering of alpha particles when fired through gold foil, known as the Geiger-Marsden or Rutherford gold-foil experiment. It was in this experiment that the German physicist Johannes Wilhelm "Hans" Geiger (1882–1945) started to use a device that he and Rutherford had invented to painstakingly count, in the dark, the scintillations recorded on a zinc sulfide–coated paper screen as it was hit with alpha particles. The instrument is now called the

Geiger or Geiger-Müller counter (named after his collaborator, the German physicist Walther Müller).

In their research, when the alpha particles were fired through the gold foil, most of the particles went right through. Some of them, however, were deflected in a rare manner that clearly appeared to be a direct boomerang back to the source, as if the positively charged alpha particles had hit a very hard object. It was "almost as incredible as if you fired a 15-inch shell at a piece of tissue paper and it came back to hit you," Rutherford said of the head-scratching phenomenon.

Rutherford intensely struggled intellectually and labored hard experimentally to further clarify, prove, and make sense of the different patterns of deflection of a few alpha particles, in contrast to the total penetration and transmission of most of them. The payoff was an atomic structure with a nucleus. As was implied by the peculiar deflection pattern, most of the positive charge was concentrated in a core that some alpha particles bounced off from in varied angles of deflection, depending on whether they hit the spherical nucleus directly in the center or anywhere in the periphery. The rest of the atom was mostly empty space that allowed the bulk of alpha particles to sail right through.

Rutherford constructed an atomic model with a small, dense, positively charged central core containing most of the atomic mass in the middle of a vast empty space. Then he had the negatively charged electrons orbit around this nucleus with wide empty spaces in between them. Democritus's indivisible atom was now proven to possess two parts.

Rutherford's atom had a different structure from Thomson's plum-in-a-pudding version, which very soon bit the dust. Introduced in 1911, this finding conferred on him the title of "father of nuclear physics" and ensured his elevated stature in science. Rutherford's and his collaborators' extensive experiments with alpha particles and nitrogen atoms led to the first-ever splitting of the atom in 1917, in the first artificially induced nuclear reaction. This quantum leap in knowledge and ability would ultimately launch the era of atomic bombs and nuclear power reactors.

Continuation of the same experiment brought about his discovery of the positively charged proton, which he named: yet another of the important contributions to science that seemed to easily and habitually come out of his own work and the work he did with others.

Bringing with him a reputation for supporting his colleagues in Montreal and Manchester, he took over the directorship of the Cavendish Laboratory in Cambridge in 1919. He enthusiastically encouraged the ideas and dedicatedly supervised the work of everyone under his watch. His laboratory churned out numerous significant contributions to science and helped create several famous Nobel laureates.

Rutherford's colleague James Chadwick claimed the 1932 Nobel Prize for his discovery of the neutron a decade after Rutherford had theorized its presence in the nucleus in order to neutralize the repulsion between same-charge protons. In that same year, his young students John Cockcroft and Ernest Walton performed the first nucleus-splitting experiment using the first-ever particle accelerator they had constructed on their own. Jointly, they won the 1951 Nobel Prize in Physics for their pioneering effort. It is unbelievable how much knowledge of the atomic nucleus was generated in a short span of time from one man and a few colleagues within his immediate influence.

THE RUTHERFORD EFFECT REACHES DENMARK

A young Dane by the name of Niels Bohr (1885–1962), a budding theoretical physicist at the University of Copenhagen and an ardent admirer of Ernest Rutherford, embraced the nuclear atom with gusto, as he did Max Planck's energy quanta and Einstein's quantum of light (although at first, he did not believe in Einstein's photons). He suspected the nuclear atom heralded the demise of Newtonian classical "gross" physics in the world of tiny matter. But there were basic problems with the structure of the Rutherford model. Notably, the model did not specify how the negatively charged electrons were positioned within the atom. If they remained stationary around the nucleus, then the positively charged nucleus would surely attract the electrons toward it. If they were moving in orbit around the nucleus, then they would accelerate and go crashing down into the nucleus, according to the laws of motion established by Newton. The negative electron charge itself would lead to instability, as the electrons would repel one another while in orbit.

To complicate matters further, the law of electromagnetism according to Maxwell's equations dictated a continuous loss of energy from electromagnetic radiation during orbit, which would add to the predicted spiraling of the electrons down toward the nucleus. The suspected instability of the electron orbit evoked serious questions on the validity of Rutherford's nuclear atomic model. Physicists recoiled.

Bohr was known for his uncanny ability to spot the flaws of other scientists' postulates and to painstakingly and critically analyze them until he clearly grasped a particular problem or was led to pursue a new project based on knowledge gained from his analysis. The very nature of Rutherford's model intrigued him and spurred him to work on it further. Bohr trusted Rutherford's experimental prowess and sound judgment so unconditionally that he was sure he could find the underlying laws that made his idol's atom tick. He suspected that the laws he needed to search for were those that rule the quantum realm. After all, he had just heartily accepted the quantum that Planck introduced and Einstein advocated. The new physics fascinated him from the get-go.

It was during this search that Bohr found the work of John Nicholson (1881–1955), an English physicist and mathematician. According to Nicholson, the momentum of an object moving in a straight line may be determined by multiplying its mass by its velocity. If the object was orbiting in a circle, however, then the circle's radius would affect its momentum and trajectory. In other words, the orbiting electron has angular momentum. Following the rule set by angular momentum, an electron has to be in a specific orbit and should not aimlessly whiz around the nucleus. Staying strictly in its orbit would allow it to circle the nucleus endlessly without ever radiating energy.

In Nicholson's model, the atom has a specific number of orbits according to the number of electrons it possesses. In this scenario, the electron has to oscillate in the orbit that corresponds to its specific frequency of vibration. To Bohr, this electron frequency limitation indicated that the electron orbit itself is quantized.

A decade later, young Prince Louis-Victor-Pierre-Raymond de Broglie (1892–1987), an up-and-coming brilliant French physicist, would posit that the electron exists in a wave-particle duality and orbits the nucleus as a stable standing wave of a specific frequency (discussed at length in chapter 5).

For some time, scientists had been noticing spectral lines or a series of colors formed by a beam of light, appearing as either discrete bright-colored lines on a dark background or dark lines on a bright-colored background, now considered to be the fingerprint of an atom or a molecule, a phenomenon they could not explain then. But to Bohr, who restricted radiation from stationary electron orbits, it was clear. Bohr happened to also be studying the work of Johann Balmer (1825–98), a Swiss mathematician who predicted a series of light spectral lines for hydrogen in the infrared and in the ultraviolet frequency ranges through a certain formula of numbers that not everybody understood very well. Balmer's hydrogen spectral lines gave him a clue to the puzzle.

An atom is stable and in "ground state" when its electrons are in their lowest energy or frequency levels. For example, the single electron of a hydrogen atom in ground state occupies the lowest-energy orbit closest to the nucleus, where it is stable. When the hydrogen electron absorbs energy, it assumes an excited state and jumps instantaneously to a higher frequency orbit—an unstable position. This excited electron has to jump back to its lowest-energy orbit, radiating its excess energy as light in the process. The amount of energy emitted is equivalent to the difference in electron frequency between the orbits, which accounts for the different spectra of light radiation. Bohr identified this difference in radiation frequency as being responsible for Balmer's bright spectral lines. (Each atom, element, or molecule is characterized by a specific spectrum of light radiation. Every spectrum is unique and characterizes only one object, just as a human fingerprint identifies only one person.)

The clarification jived well with Planck's formula $E = hv$, where E is energy equal to Planck constant h times frequency v. Bohr's atom absorbed and emitted light in quanta of energy. This was exactly the law Bohr was looking for to support his faith in the stability of Rutherford's atom.

An electron in quantized orbit emits radiation only when it jumps to a lower frequency orbit and releases energy. Until such time, it remains in a fixed orbit. Then a question popped up in Bohr's mind. How does the electron know its destination so it could emit the specific frequency when it jumps down? He had no answer at the time.

Two years after Rutherford introduced his nuclear atom, Bohr published the first paper of his trilogy on the quantum atom. For the perfectionist Bohr, this was an agonizing passage through a difficult maze riddled with obstacles posed by the limitations of entrenched classical physics. The process was made even more challenging by Rutherford, to whom he had submitted the papers, not only because senior physicists had to endorse reports written by junior colleagues to ensure their speedy publication but also because he very much valued his hero's opinion.

His pain was eased some by his fiancée, Margrethe Norlunde, who would become his wife as he grappled with his dilemma. She jotted down his ramblings and corrected his faulty English as he paced around, collecting and verbalizing his thoughts—essentially becoming his secretary out of necessity.

Rutherford found Bohr's radical and novel ideas on the behavior of the electron difficult to accept, much less to visualize, and his presentation too long and circuitous. He offered to shorten his trilogy of papers and to correct Bohr's choice of words, to no avail. For the Dane, who was notorious for littering his study with discarded drafts, the thought of having some of his carefully chosen words and his meticulously woven concepts edited from the document was so repulsive that he went to personally see Rutherford in Manchester. He doggedly defended his paper until his idol relented in exhaustion.

The trilogy was published in July, November, and December 1913, entirely unabridged. The new innovative and revolutionary atomic model it presented was thoroughly studied and critically discussed by scientists worldwide and was found to finally provide the elusive explanations for the many observed phenomena that had long bewildered them.

The reception of the quantum atom was robust in all different directions. Bohr's quantized atom was wild, daringly bold, and challenging. It incited heated discussions and evoked equally strong positive as well as negative emotional reactions among his peers. Classical physics was so deeply ingrained, that uprooting it from the intellects and hearts of those who had lived and worked with it for so long was a difficult and painful undertaking.

The rejection on one side was dripping with disbelief and despair. Many physicists could not accept Bohr's quantum atom model's violation of Maxwell's electromagnetism equations, which dictate that all electrons moving in orbit must emit radiation. English scientist Sir John Rayleigh could not believe that "nature behaved this way." On the other side, there was Einstein, who pronounced the model as "one of the greatest discoveries," and English physicist and astronomer Sir James Jeans, who defended it as "needing no more justification than its success."

Bohr had provided an atomic model with a solid theoretical foundation for the stability of Rutherford's nuclear atom that was nothing short of brilliant, and its unmasking of the mystery of spectral lines was stellar. The model's quantization of energy radiation justified Einstein's photons. It was vigorously tested in experimental laboratories, where its predictions were verified and newer utilizations found. Bohr's quantum atom survived many rigorous scientific tests. It steadily achieved acceptance in the community and grew in scientific application.

Bohr received the Nobel Prize in Physics in 1922 "for his services in the investigation of the structure of atoms and of the radiation emanating from them."

THE ATOMIC STRUCTURE

The following description of the atomic structure has to be taken in a historical context, as it has changed with the accumulating discoveries of its nature. What's known about it now might still change as science uncovers its hidden attributes and as scientific language evolves to better describe its physical characteristics.

The atom has been computed to be roughly a hundred thousand times larger than its nucleus, which, in turn, is two thousand times larger than the electron orbiting around it. The nucleus consists of positively charged protons and zero-charged neutrons surrounded by negatively charged electrons. Quarks are subatomic elementary particles classified into six "flavors" of up, down, top, bottom, strange, and charm—the complexities of which are beyond the scope of this book. Two up quarks and one down quark constitute a proton, one up quark and two down quarks, a neutron; all of them tightly bound by the "strong nuclear force" mediated by the gluon. The weaker attractive power of the "weak nuclear force" carried by the W and Z bosons allows change in the combinations of quarks. For example, a neutron can change into a proton which upsets the neutron-proton ratio. This loss of balance decreases the binding strength of the nuclear strong force, making the nucleus unstable. The nucleus undergoes radioactive decay, releasing a beta particle, an electron and an anti-neutrino (a particle with no mass but has energy and momentum). The process ignites the fusion of two or more nuclei, releasing vast amounts of energy responsible for sustaining the stars and most of the light we see.

The neutrons serve to stabilize the protons, which, having the same positive electric charge, would otherwise repel one another (this repulsive force which pushes the protons apart is called the electromagnetic force). Every proton is matched by a negatively charged electron located in a specific-frequency "orbital" (the term for the orbit of a wave-like electron). When the numbers of protons and electrons are equal, a balanced or neutral atomic charge is maintained, an optimum state for the human body.

The number of protons in the nucleus determines the atomic number of that particular atom, and the protons, neutrons, and electrons together make up the atomic mass. The atomic weight is controlled by the pull of gravity on the atom. Traditionally, atomic mass and atomic weight were used interchangeably, since their numbers are equal on earth. Away from the earth's gravitational pull, however, they part ways. Astronauts are almost weightless on the moon, as the gravity of the moon is only 17 percent of the earth's gravity; in contrast, their atomic mass remains the same, since their atoms still have the same number of protons, neutrons, and electrons.

The internal structure of the electron has yet to be formally described. Some scientists argue for a constant flow of energy in the shape of a double torus, a mathematical structure resembling a doughnut.[3] The toroidal center may be likened to an energy-packed doughnut hole. Electrons are now said to exist as waves of probability propagating through space, in which the square of the magnitude of the wave at any given point yields the probability of finding the particle at that point (discussed in detail later in this chapter). The orbitals of the electron are harmonics of the electron's probability waves arranged around the nucleus in a specific mathematical order. This is the answer to Bohr's question of how the electron knows exactly what orbit it should jump to.

Moreover, the electron does not travel around its orbital as a solid object. Even if it were technologically feasible to do so, one cannot take a video of an electron zipping swiftly around the nucleus. It is nowhere and yet everywhere, existing purely as waves of vibration, only as energy potentialities for particulate material existence.

When an electron absorbs photon energy, it gets excited and jumps up to a higher frequency orbital, following the mathematics of harmonics. When the electron gives up this energy, its frequency changes to a lower number, and it jumps down to a lower-energy orbital. This jumping up or down is called a "quantum leap" and is the principle behind light-emitting diodes (LEDs), which make possible thin television screens, remote controls, digital clocks, long-lasting and energy-efficient light bulbs, and many other electronic devices.

The term has found a niche in common parlance in our world of sticks and stones. It sounds so extraordinarily bold and quickly decisive that we use it to aptly describe an abrupt, major, and

---

[3] See Ginzburg (1998) and Correa and Correa (2011) in the bibliography.

earthshaking change of a life circumstance, belief, habit, or paradigm long held, usually with an attendant strong emotional flavor. There probably is no other phrase that expresses a jarring life event better than a "quantum leap."

The course of the electron's transition to a different-frequency orbital is definite and instantaneous, not a slow meandering up or down the spectral gap to the orbital of destination. It is either this frequency or that frequency. The choice is clear-cut.

The number of electrons determines the atom's chemical behavior. In Bohr's atom, the outermost electrons are called the "valence shell" and are involved in bonding with other atoms. It is in this valence shell where chemistry starts and biology takes off. Through the interaction between valence electron vibrational frequencies, atoms become molecules and molecules become cells and cells become organs and organs finally become organisms, ranging from plants to humans and our loyal friend, Fido the dog.

The rich designs and varied architecture of all objects found in nature are the end results of combining electrons of different atoms, some of them sharing one electron between orbitals of two or more atoms, thus locking the molecule into a stable harmonic composition. The resonant bonding of electrons is the secret behind the "specific properties" of Democritus's atom that guides its interactions to smoothly and beautifully shape one's reality.

The addition of protons and neutrons to the nucleus of the simple hydrogen atom, along with their electrons generates the list of elements populating the periodic table that we learned in high school chemistry. The elements vibrate, each one creating a symphony of frequencies expressly characterizing the atoms and molecules that constitute that particular element.

This attribute accounts for the specific combinations of spectral lines of radiation from an element or an object, or a human body, that are its very own fingerprint. The new classification of the elements is now based on their frequencies and harmonics.

Fast-forwarding to the present, the still-invisible atom is indivisible no more. The atom seems to be divisible to infinity, as if nothing is there whatsoever. If you take a very high-powered microscope such as the Large Hadron Collider (LHC)—which can measure within a millionth of a millionth of a millionth of a meter—and look into a grain of sand, you'll see that it is made of intricately structured crystal lattices of molecules looking like fine lace, translucent and delicate. The sand you are looking at is quartz, the crystal that reflects back all colors of the rainbow and does not absorb much heat. Quartz is what gives the powdery sand on Florida's Siesta Key beaches their gleaming whiteness and barefoot-friendly warmth.

As you gradually increase the magnification, the delicate lace-like structure disappears, and what you see are mainly big holes in between the lattices. Progressive magnification results in the same disappearing act of the structure you are focusing on. Whatever you are looking at simply melts away on closer inspection. Where is the atom? Where is the nucleus that Rutherford's alpha particles bounced against? There is nothing there, it seems, except an expansive sea of energy in a ceaseless rhythmic dance of creation and dissolution. Pure energy vibration is all there is.

The penetration into the nature of the atom has progressed tremendously since Bohr introduced his quantum atom model. Still, what we know of the elements of the world of the very tiny is based on laboratory observations of the results of interactions between these particles and of their by-products, and not from direct observation of quantum particles and quantum operations. The electron has not been observed jumping from a high-energy-state orbital to a lower energy orbital; it is only the spectral lines of light produced by released energy in the form photons that have been successfully recorded. The Higgs boson, a particle named after the British physicist Peter W. Higgs (b. 1929) and dubbed the God particle, was not what was observed at the Large Hadron Collider at CERN (Conseil européen pour la recherche nucléaire, or the European Organization for Nuclear Research) in Geneva, Switzerland. The Higgs boson production from high-speed proton-to-proton collision at the facility's 27-kilometer-long (16.7-mile) particle-accelerator ring was deduced after its mathematically predicted distinguishing characteristics were calculated from the by-products of proton decay. These by-products included two photons, the particular characteristics of which gave away one of the major predictions for the Higgs boson: its calculated mass of 125 GeV (giga-electron-volts).

Regardless, this knowledge arrived at by high-speed collisions is a far cry from that gained from direct impact on the senses. There is a dearth of descriptive expressions in this dry new world of particle collisions and decay. Since language evolved from our need to impart concepts formed by sensory inputs—such as those aroused by beautiful music, hair-raising adventure, a riveting book, silky smooth skin, a delicious dinner, and itchy feet—the words we know fall short of fully expressing the quantum. New terms have to be coined to convey the unfamiliar experience.

Indeed, the paucity of eloquence in this area was articulated by Ludwig Boltzmann: "How awkward is the human mind in divining the nature of things, forsaken by the analogy of what we see and touch directly."

5
The Quantum March

We are given to understand that ... that mysterious boundary between the
subject and the object has broken down.
Erwin Schrödinger

Max Planck's light-energy quanta, Einstein's quantized light photons, and Niels Bohr's quantized electron orbit firmly laid the first foundations of quantum physics. What happened thereafter was nothing short of an inexorable advance in theoretical speculation, knowledge, research activity, and excitement in the scientific community. The phenomenal progression of the quantum pushed on so relentlessly that it begs to be called a "quantum march," a sweeping and electrifying succession of events that thrilled physicists and ordinary people alike and changed how we, modern humans, conduct our daily lives.

One discovery quickly followed another in physics, astronomy, biology, chemistry, industry, and commerce. The application of quantum mechanics—which mathematically expresses quantum theory—in industrial laboratories blossomed into better equipment for humanity, the sciences in particular. The twentieth century was a glorious time for science and industry. Computers, telescopes, transportation, navigation, space exploration, wireless communication, devices for medical diagnostics/therapeutics, and many more areas of use were affected in a way that enhanced the lives of many mortals.

The very first congress on quantum theory was the 1911 Solvay Conference held in Brussels, Belgium. Considered to be of prime importance in the fields of chemistry and physics, it was started by the Belgian industrial chemist Ernest Solvay, who, heeding the advice of Hendrik Antoon Lorentz, a famous Nobel prize-winning Dutch physicist, and Walter Nernst, a German chemist (who won a Nobel prize for his third law of thermodynamics), made a fortune from his soda ash (sodium carbonate) production system.

Lorentz chaired the weeklong summit in a no-holds-barred affair at the elegant and luxurious Hotel Metropole, marked by a rare level of hospitality that solicitously catered to the needs of the by-invitation-only attendees. Even their travel expenses were covered. The thirty-two-year-old Einstein was the youngest invitee, but the youngest attendee by far was nineteen-year-old Louis de Broglie, who accompanied the scientific secretary of the conference, Maurice de Broglie, who just happened to be his older brother. The 1927 Solvay Conference would subsequently showcase this young prince as one of the rising stars of physics.

During the 1911 Solvay, Einstein was asked to present his paper on the law behind the photoelectric effect, not on the quantum nature of light itself. Quantum theory was so radically unfamiliar and so contrary to classical Newtonian physics that it confused some, irritated many, frustrated a few, and most assuredly intrigued everyone, no one more than Bohr, who embraced the quantum and its implications with such zest and relish that irked, yet at the same time stimulated Einstein. These two scientific giants' ongoing intense debate on quantum physics sparked and underscored their lively encounters and forged between them an intellectual relationship worthy of envy among their peers.

THE GOLDEN TRIANGLE

The universities of Munich and Göttingen in Germany and Copenhagen in Denmark formed the golden triangle of theoretical physics. These three major institutions became the hub of intellectual activity devoted to the study of the quantum in the early years of the twentieth century. Quantum physics was nurtured in these universities through their respective theoretical physics directors' inspiration and untiring guidance, reinforced by the freely generous sharing of ideas between highly esteemed teachers and young, energetic, and exceptionally talented students and junior colleagues.

Wilhelm Röntgen, the Nobel laureate discoverer of X-rays and director of the physics institute at the University of Munich, selected the German theoretical physicist and master educator Arnold Sommerfeld (1868–1951) for the prestigious positions of "extraordinary professor" of physics (außerordentlicher Professor in German) and director of the theoretical physics institute. At this time, theoretical physics had superseded experimental physics as the leader and mover of the new physics, driven by the highly charged quantum movement and the rise of superior theorists such as Niels Bohr, Max Born, Erwin Schrödinger, Albert Einstein, and Arnold Sommerfeld himself. Experimental physics assumed a supporting role to advance or verify quantum physics theory.

Sommerfeld was one of the early pioneers of quantum physics. Young atomic and quantum physicists such as Wolfgang Pauli (1900–58) and Werner Heisenberg (1901–76) flocked to the University of Munich to obtain their doctorate degrees or simply to learn from this consummate teacher. Like Ernest Rutherford, Sommerfeld supported the ambitions and promoted the efforts of his students, many of whom achieved fame in their own right. Several former students became Nobel laureates, a distinction that frustratingly eluded Sommerfeld even though he was nominated some eighty-four times through his lifetime. His extraordinary intellectual power and widespread popularity as a teacher and mentor were such that by the year 1928, close to one-third of the ordinarius professors (professors of the highest rank) of theoretical physics in German-speaking institutions were his former students.

The highly respected Jewish-German mathematician, physicist and science author Max Born (1882–1970) was one of the prime movers of quantum mechanics. He won the Nobel Prize for physics in 1954 for his "fundamental research in Quantum Mechanics, especially for his statistical interpretation of the wave function." When he moved from the University of Frankfurt am Main to the University of Göttingen to become the director of its physical institute, his negotiated package included the installation of long-time friend and collaborator, experimental physicist James Franck (1882–1964), as chair of experimental physics.

Born, the theoretician, and Franck, the experimentalist, shared the same views on basic scientific concepts, and their twelve-year (1921–33) successful scientific teamwork enhanced each other's career during the golden years of the quantum, when highly spirited intellectual and experimental activities were concentrated in the three institutions composing the golden triangle.

It was when Sommerfeld took a trip to the United States that Heisenberg, who was studying under him, joined Born as an assistant. He became a *privatdozent* (unsalaried lecturer) at the University of Göttingen when he completed his studies. In 1925, Born and Heisenberg jointly presented quantum theory in the mathematical language of matrix mechanics, effectively enhancing understanding of quantum mechanics and easing its acceptability.

Follow-up papers by Heisenberg, Born, and the physics whiz kid Pascual Jordan (1902–80) established quantum mechanics on a solid mathematical foundation. In 1928, Einstein nominated the three collaborators for the Nobel Prize in Physics. Curiously, the committee selected Heisenberg alone for the 1932 Nobel Prize, "for the creation of quantum mechanics, the application of which has led to the discovery of the allotropic forms of hydrogen."

The sad and awkward situation tore young Heisenberg apart. He thought the work had been a joint effort, and he felt that all three of them should have shared the coveted prize.

The University of Copenhagen established the chair for theoretical physics especially for Niels Bohr. Opened in 1921, the Niels Bohr Institute was patterned after the way Rutherford ran his laboratories, including his approach toward successful mentorship and collaboration with students and junior colleagues. Scientists from all continents came to learn from and share ideas with the physics master for whom the institute was named. The bulk of the effort expended on scrutinizing the atom and the quantum was done in this institute, with help from other physicists at different institutions. It was at this institute that the famous "Copenhagen interpretation" of quantum theory (discussed later in this chapter) was introduced. It was from here that Werner Heisenberg perplexed and charmed the world of science with his uncertainty principle. It was in this institution that quantum theory and the understanding of it developed—largely from the work of Bohr and Heisenberg, with contributions from Born at the University of Göttingen.

TEARS OF SCIENTIFIC JOY

"It's so lovely, it makes you cry!" Einstein remarked after Franck presented the first experimental confirmation of atomic energy quantization, a joint effort with the German physicist Gustav Ludwig Hertz (1887–1975) that led to their 1925 Nobel Prize in Physics. (Gustav Hertz is not to be confused with his uncle, Heinrich Rudolf Hertz, for whom the hertz unit of measurement is named.) They showed that a boost of electricity can force an atom into an excited state, thus confirming the quantum prediction that electrons exist in discrete quantized energy states—a requisite in Bohr's atomic model.

It was already accepted that when an electron leaps from a higher to a lower energy level, the electron spontaneously emits a quantum of light radiation. This spontaneous emission of radiation is responsible for most of the light we see, sunlight itself mostly being a spontaneous emission of energy radiation from the biggest blackbody in our neighborhood, the sun. The glowing embers in the fireplace, a friend's warm hands, and a young child's face, red from playing in the summer sun, are all due to the spontaneous release of heat and light by leaping electrons.

Building on Franck and Hertz's intentional electrical excitation of the atom around the phenomenon of spontaneous emission, Einstein employed a new and breathtaking twist. In a freshly different utilization of the quantum leap, he deliberately had an electron absorb photon energy and jump to a higher-energy-level orbital. He then argued that when the fully excited electron, which was now incapable of absorbing any more energy, was again intentionally hit with a photon of a specific frequency, the extra and redundant energy would eject the fully excited electron from the higher orbital and induce the electron to leap down to the lowest energy level, resulting in the release of radiation. He called this "stimulated emission."

This was a clever extension of the photoelectric effect. It was cutting edge. Still, Einstein was not happy with what he had uncovered about nature's behavior in the world of the tiny. He just could not explain why stimulated emission was unpredictable in its time of occurrence and its direction of radiation. The consistency of cause and effect of classical thought was absent in the realm of the quantum atom. Einstein was moved to say, "I find the idea quite intolerable that an electron exposed to radiation should choose of its own free will, not only its moment to jump off, but also its direction. In that case, I would rather be a cobbler, or even an employee in a gaming house, than a physicist."

Einstein's unhappiness over the puzzling stimulated emission was sharply contradicted by the enormous thrill it ignited among industrial physicists. Stimulated emission was developed further by several scientists, including Richard Gordon Gould (1920–

2005, a pioneering American physicist at Columbia University in New York. In 1957, he built a device that delivered a narrow and intense beam of light, which he named a laser.[4] In 1960, Theodore Maiman of the Hughes Research Lab in Malibu, California, made a functional model of the laser, and the rest is history. Today, mere decades after the idea was introduced, lasers are applied in pointers, bar code readers, compact disc players, laser printers, fiber optics, entertainment, search lighting, laser cooling, metal cutting, welding, and medical research, diagnostics and treatment, including surgical operations. Once more, Einstein's incredible genius flames excitement in an all-too-ordinary world.

Nevertheless, it is the single-frequency monochromatic beam laser technology that has found its way to a position of particular importance in our daily lives. The pure light beam gives the laser spatial coherence, which indicates that it can be focused on a very sharply defined spot, making it desirable for fine and accurate cutting: in surgical procedures in medicine (in all organs in general, eye, blood vessels, skin conditions and hair removal); dental; veterinary; industrial (e.g. metal sheet cutting); and high-precision measurement. The laser's single-color emission allows it to be pulsed in short bursts in orders of femtoseconds (one femtosecond equals one quadrillionth of a second). Such lasers are used in radar devices[5] to track ballistic missiles and detect geological cracks and invisible damage to aircraft skin, motor vehicles, and ships; and in target marking and astronomy.

With today's advanced laser technology, the previously thought unruly behavior of the stimulated electron is easily and completely controllable, following physical law. Einstein did not need to wish to be a cobbler or a gaming-house employee.

---

[4] The term "laser" stands for "light amplification by stimulated emission of radiation."

[5] "Radar" stands for "radio detection and ranging."

A MATTER OF PARTICLES AND WAVES

Louis de Broglie, the young French prince who had finagled his attendance at the first Solvay Conference in Brussels through his brother, brooded over the dual nature of light as both particles (photons) and as electromagnetic waves, and he wondered if this phenomenon could be extended beyond light photons to other particles—that is, the material contents of the universe. Could electron particles likewise behave like waves?

Assuming a positive answer to the question, he proceeded to designate specific wavelengths and frequencies to electrons. He assigned them only to orbits that could accommodate a whole number count (no fractions) of wavelengths so that a continuous series of waves around the nucleus would form an exact fit. By so doing, he discovered that a "standing wave" had developed around the atomic nucleus, a phenomenon not feasible with wavelengths that did not have the correct whole number (or fractionated) to allow a perfect fit of the waves.

A standing wave results when two same-frequency waves that are traveling in opposite directions arrive at a certain point at the same time and constructively interfere with each other. Since the two waves are of identical frequency, they do not travel past one another. They just vibrate together, as though they are one, their combined and amplified wave crests and troughs appearing to be just oscillating in place and going nowhere, the very reason for its name. A standing wave is stable and still.

He also found that Bohr's quantum number n, assigned to the orbit the hydrogen electron was restricted to, happened to correspond exclusively to the orbit permitting the existence of a standing wave. The electron had to circle the nucleus as a stable standing wave of a specific wavelength and frequency, without the propagation of wave motion and without the particle itself going around the nucleus (figure 4). Thus, there would be no energy loss to send the electron spiraling down the nucleus. De Broglie showed that the electron definitely existed in a wave-particle duality. In a single stroke, the prince handed Bohr an impeccably elegant rationale for his quantized electron orbits and for why electrons did not radiate during orbit.

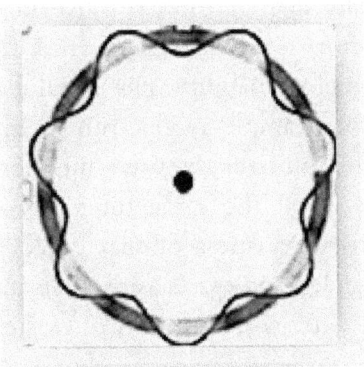

Fig. 4. The de Broglie electron standing wave around the nucleus. The ring represents the electron's phase as frequency, and the wavy line shows the seven whole-number wavelengths around the nucleus. Source: Pieter Kuiper. Public domain, Wikimedia Commons.

A MATTER OF WAVE FUNCTION

Erwin Schrödinger (1887–1961), an influential Austrian physicist and philosopher, was working on a universal quantum equation that would extend Newton's laws of motion governing large objects such as balls and bats to the tiny quantum atoms and electrons he was dealing with. De Broglie's wave nature of the electron was a more appealing idea to him than Bohr's orbiting electron particle, which radiated only when the electron suddenly jumped from one orbit of higher energy to another of lower energy in a quantum leap that did not make any sense to him. He tackled this problem while on a productive and inspired December

1925 holiday with a female friend at a Swiss alpine mountain resort and during the following six months of exhaustive mental activity. The fruit of his labor and romantic inspiration is the now-famous equation that became the foundation for wave mechanics.

Schrödinger's wave function equation represents matter of any size moving at any speed, thereby covering the description of objects ranging from tiny particles such as electrons to baseballs and stars as waves of probability. His initial explanation was that a material object, for example an electron, is "smeared" as vibratory waves of probability all over the wave function across all of space. Wave function is all at the same time the object itself and the spread-out vibratory waviness of that object. If the object is an electron, then the electron exists as particle and wave together, all dispersed about as energy potentiality of existence.

The particle and the waves of probability do not have separate identities until an act of observation or measurement brings about the "collapse" of the wave function. (The terms "reduction" and "collapse" have been used interchangeably in referring to this wave function phenomenon.) This collapse reduces or forces the waves of multiple probabilities into a defined energy state and compels the electron to crystallize into a particle where it has the greatest probability of manifestation.

Furthermore, it cannot be said that only a major bulk of the electron matter is found in the area where it is observed at the collapse of the probability wave function and a little part remains elsewhere. That is not the case. When one sees the electron in any one spot, it is either wholly and 100 percent there, with no chance for a part of it to be somewhere else, or wholly and 100 percent not there at all.

The smeared wave-particle state of matter was not a pleasant image to envision and was unpalatable to many physicists. The major development of wave function theory that made it acceptable was an accomplishment by Max Born at the University of Göttingen. Working alone, he formulated an interpretation of the "probability density function" for Schrödinger's equation. He clarified that the wave function does not describe a particle's behavior; rather, it expresses the statistical probability of a particle being found in a specific location at a particular time imbued with specific characteristics, depending on the amount or density of energy present in those waves of probability. In other words, a crucial energy strength is required to drive the wave-particle's manifestation into a definite particulate material reality.

Published in 1926, the work is now accepted as the standard interpretation of Schrödinger's most successful wave function equation. He shared the 1933 Physics Nobel Prize with English physicist Paul Adrien Maurice Dirac (1902–84), who predicted the existence of antiparticles, "for the discovery of new productive forms of atomic theory."

THE WONDER BOYS OF PHYSICS

The twenty-four-year-old Werner Heisenberg, a much-admired whiz of physics, had just published an article the summer before Schrödinger's productive alpine holiday. The paper outlined a mathematical foundation for quantum mechanics, with its matrix calculus describing particles. Heisenberg formulated the work during his vacation in Helgoland, a red sandstone island peeking from the sea about fifty kilometers (thirty miles) from the German mainland, an ideal place for him to find respite from a terrible bout of hay fever.

Schrödinger did not like Heisenberg's approach and vice versa, although both agreed on the soundness of their individual mathematics. Heisenberg would later acknowledge that the two approaches amounted to exactly the same thing, only presented differently. Heisenberg's matrix mechanics describes particles, while Schrödinger's wave function describes vibrations. They both refer to a single reality that mystically exists in a wave-particle duality. Schrödinger's wave function approach was much easier for physicists to apply, however, and is the equation scientists prefer to use today.

Heisenberg was only eighteen years old when he enrolled at the physics institute headed by Sommerfeld at the University of Göttingen. During the first seminar he attended there, Heisenberg made friends with Wolfgang Pauli, a Viennese physics wonder boy who was widely considered to be the most gifted of Sommerfeld's students. Pauli became a significant influence on Heisenberg's professional life: first, in steering him toward quantum mechanics, and second, in being a sounding board throughout his intellectual struggles with the bothersome quantum.

The second person Heisenberg met in Göttingen who was of utmost importance to his career was Niels Bohr of the University of Copenhagen. In one of Bohr's lectures as a visiting professor at the University of Gottingen, Heisenberg called Bohr's attention to a few difficulties in a publication they were discussing. This must have impressed Bohr, since the professor asked Heisenberg to take a stroll with him after the lecture: a vivid memory in the student's youth that he dearly treasured. He was subsequently offered a chance to study with the quantum master in Copenhagen.

When Heisenberg arrived at Bohr's institute, Pauli had already finished his stint and had moved back to Hamburg, Germany. The problem badgering the physics community during this time was the splitting of the spectral lines of light radiation of an atom in a magnetic field, called the "anomalous" Zeeman effect (named for the Dutch physicist Pieter Zeeman), a mystery that consumed Pauli and caused him much despair. He hoped that Heisenberg could solve the problem. The two exchanged ideas during Pauli's frequent visits to Copenhagen, a train ride away from Hamburg. They also communicated often by mail. It was with Pauli that Heisenberg shared his thoughts about matrix quantum mechanics.

EXCLUSION BY SPIN

An electron was afforded a certain degree of freedom within the atomic structure outlined by Bohr: it could move up and down, left and right, and forward and backward. Each degree of freedom was assigned a quantum number—n, k, and m—and each energy state or electron orbit was defined by the series of combinations of the three. This was the accepted state of affairs, except for an anomalous splitting of the atomic spectral lines to more than what could be accounted for by three numbers and their combinations (i.e., the Zeeman effect).

Pauli agonized over this problem until he hit upon the idea of an additional degree of freedom, a fourth quantum number, later identified as "spin,"[6] that was allowed two values (spin one-half, up or down), which doubled the quantum number combinations and thus, the number of energy states. This new approach made sense of the observation that filling the electron shell required twice the number of the then-known available electrons.

This finer dissection of the atom led to the "Pauli exclusion principle," which states that no two electrons in an atom can have the same set of four numbers. Electrons are prevented from bunching together in the electron shell at the lowest energy level. This new model set Bohr's quantum atom on firmer ground, eased the introduction of the up or down electron spin, and eventually facilitated the establishment of quantum mechanics, of which spin is an essential component. Pauli was given the 1945 Physics Nobel Prize for the discovery of his exclusion principle.

---

[6] According to Victor J. Stenger (1935–2014), an American physicist, philosopher, and author, "spin is the total angular momentum or intrinsic angular momentum of a body. The spins of elementary particles are analogous to the spins of macroscopic bodies. In fact, the spin of a planet is the sum of the spins and the orbital angular momenta of all its elementary particles. So are the spins of other composite objects such as atoms, atomic nuclei and protons (which are made of quarks)."

Electron spin was not Pauli's idea, although it did support his two-valued fourth quantum number theory. It was a twist on the said theory that occurred to Ralph Kronig (1904–95), a twenty-one-year-old German-American whose curiosity was aroused after he had read Pauli's letter about the fourth quantum number to Alfred Landé (1888–1976), whom he was visiting in Frankfurt in January 1925. (Landé was a German-American quantum theorist who provided an answer for the Zeeman effect.) Kronig worked on the theory's mathematics and discussed them with his host before Pauli's arrival the next day. Pauli was coming specifically to discuss the exclusion principle with Landé, whose opinion he held in great esteem. Both Kronig and Landé thought that quantum electron spin was a brilliant explanation of the extra degrees of freedom, and they were utterly dumbfounded when Pauli rejected it offhand, saying: "It is indeed very, very clever but of course has nothing to do with reality." Because Pauli was widely recognized and respected for his scientific brilliance, his word carried a lot of weight and emotionally affected Kronig. Utterly disheartened, the latter trashed his idea, thankfully not before he had discussed it with Bohr at the University of Copenhagen, where he spent the next ten months after his fateful visit with Landé and Pauli.

Kronig would soon regret this reliance on the opinion of an expert instead of his own intuition. Only ten months later, Samuel Goudsmit and George Uhlenbeck, both in their twenties and earning their physics doctorates at Leiden University in the Netherlands, published their thesis on electron spin in which they presented the mathematics that Kronig had mostly figured out before Pauli's arrival. Both doctoral candidates had no prior knowledge of Kronig's theory.

Disgusted over having missed out on electron-spin because of Pauli's outright derision of the idea and for allowing Pauli to have undue influence on himself, Kronig shared his frustration with Bohr's assistant. Somehow, Goudsmit and Uhlenbeck got wind of what had transpired between the two scientists, and they credited Kronig with having developed the theory before they did.

Paul Dirac was a novice twenty-three-year-old research student at Cambridge University when all that agitation was taking place. On arriving in Cambridge, he was immediately thrust into the unfamiliar world of atomic physics of Ernest Rutherford, the father-in-law of the British physicist and chemist Ralph Fowler, his supervisor. Quiet and a loner by preference, Dirac took full advantage of the time for independent study the research program allowed him and spent much of it on Niels Bohr's quantum atom.

In his effort to find quantum equations to explain the motion of electrons, Dirac's solutions predicted the existence of a particle exactly like the negatively charged electron but with a positive charge and a mirror-image spin: an antiparticle. This prediction was readily confirmed with the discovery of the positively charged antimatter positron (e+) by Carl David Anderson (1905–1991), who received the 1936 Physics Nobel Prize for it.

When a positron meets an electron, it results in their mutual annihilation and release of energy. Dirac was awarded the 1933 Nobel Prize, which he shared with Schrödinger, "for the discovery of new productive forms of atomic theory."

THE GOLDEN YEARS OF PHYSICS

The twentieth century witnessed a widespread flurry of activity and giddy excitement in theoretical and experimental physics. The same was true in industry as well, fueled by the fast and copious advances in knowledge, the application of quantum mechanics, and the generous sharing of ideas within the pioneering quantum physics community. The years between 1924 and 1927 saw a massive burst of breakthroughs from the high-energy activities of young physicists who were inspired by the prolific scientific output of Bohr and the theoretical genius of Einstein.

The breakthroughs were further fomented by the two scientific giants' passionately opposing views on the direction of quantum mechanics: Einstein desperately clinging to preserve the causality and strict determinism of classical physics and Bohr vigorously pushing for acceptance of the probabilistic nature of the quantum. Their profound philosophical chasm would invariably charge their highly animated and friendly debates throughout their lives.

As told by the author Walter Isaacson in his 2008 book *Einstein*, after receiving the 1922 Nobel Prize, Einstein went to see Bohr in Copenhagen. Bohr met him at the train station, and they started one of their animated conversations on the streetcar ride to Bohr's home. They got so engrossed in their dialogue that they missed their stop by a long way, and so they rode the streetcar back, only to miss their stop again and again. "I can well imagine what the people thought about us," said Bohr of their distracted behavior.

QUANTUM UNCERTAINTIES AND PROBABILITIES

Their intense discussion was a reflection of the confusion brought about by the deep-seated reluctance of physicists, exemplified by Einstein, to let go of Newtonian determinism and causality on one hand, and the exciting arena of quantum uncertainties, whose equations reveal true predictive power, on the other, even while negating direct or indirect "local" interactions and violating the speed of light at the same time. The causality and determinism of classical mechanics rigidly require inviolability of the speed of light and locality. (The idea of locality, now essentially discredited by the findings of quantum physics, holds that objects are affected by their environment only if they have a way, be it direct or indirect, of communicating with the environment.)

The state of uncertainty of quantum phenomena was aggravated by the introduction of Heisenberg's "indeterminacy principle," or "uncertainty principle," which is fundamental to Schrödinger's wave function equations, and which extends to other quantum systems as well. The Heisenberg uncertainty principle posits that when subjected to the smallest possible probe, the electron's features cannot be accurately determined. One may measure either its momentum or its position at any one time, but not both at the same time with precision.

This characteristic of a particle is inherent in its dual nature of existence as both a wave and a particle. Once one has identified the electron in a specific position, its wave function has already collapsed into being a particle, and one determines the electron as being there, in that fixed position. Its velocity and trajectory are no longer measurable. This is like a ball that is frozen in midair and is no longer in flight; measurement of its speed and its flight trajectory is then out of the question. The same is true the other way around: the ball's exact location cannot be pinpointed while its speed is being measured in flight.

The scientific community needed a clearer interpretation and better understanding of quantum mechanics, which is how the Bohr-Heisenberg interpretation entered the scene. Later dubbed the Copenhagen interpretation (discussed earlier), the clarification is a composite of Heisenberg's uncertainty principle, Bohr's complementarity principle and his rule of correspondence, and Born's "statistical probability density function" interpretation of Schrödinger's wave function.

The Copenhagen interpretation says that the wave function is a complete description of the wave-particle cloud of probabilities that is spread all over the expanse of space; it does not specify where in the cloud the particle is located. Only statistical probabilities of finding it somewhere exist, according to the cloud's probability-density function. The wave function strictly obeys Heisenberg's uncertainty principle. The act of observation or measurement amplifies the energy intensity of particular probability waves, and thus their statistical probability density, which in turn collapses the wave function and forces the particle to be in a specific position or to have a specific momentum or a specific attribute—as perceived by the observer.

The rule of correspondence states that although the language of classical mechanics of gross matter can describe the motions of an electron in its stable orbit, such language cannot describe the electron's transition to a different orbit and the resultant emission of radiation. The rule requires that the emission be a single-frequency radiation (which is the difference between the frequencies of the two orbitals) and that this frequency has a harmonic equivalent in the language of the classical mechanics of macroscopic objects. The rule also requires that quantum and classical calculations must yield the same answer when calculating larger quantum energies, when a huge number of electrons are jumping at the same time.

The "collapse" of Schrödinger's wave function was a product of Heisenberg's effort to provide a clearer explanation of Bohr's complementarity principle. He likened the wave-particle nature of an object to the two sides of a coin; they are simply reciprocal presentations of the same thing. Just as one doesn't see the head of a coin while looking at the tail side, one does not observe the wave aspect of an object while measuring its particle aspect. The observer gets one or the other, not both at the same time.

The collapse of wave-particle duality into material reality is in accord with the belief of an "embodied spirit," the concept of an inseparable union of body and soul found in Hinduism, Buddhism, and Christianity. The idea reminds us of a passage in the Holy Bible when the newly-resurrected Christ appeared to his grieving disciples: "Jesus Himself stood in the midst of them, and said to them, 'Peace to you.' But they were terrified and frightened, and supposed they had seen a spirit. And He said to them, 'Why are you troubled? And why do doubts arise in your hearts? Behold My hands and My feet, that it is I Myself. Handle Me and see, for a spirit does not have flesh and bones as you see I have'" (New King James Version [NKJV], Luke 24:36–39).

The Copenhagen interpretation introduced an unpreventable and uncontrollable intrinsic physical interaction between the quantum entity being measured and the classical gross measuring device that definitely entwines and affects them both; the human consciousness apparatus being the ultimate measuring device. The conscious human observer, therefore, becomes one with the object being observed and vice versa. Forever.

The Copenhagen interpretation was not popular. On analyzing Schrödinger's wave function, Heisenberg's uncertainty principle, and Born's statistical probability, Einstein protested that if a particle does not have a definite position before its measurement and that only calculable probabilities exist until it is measured, then quantum theory was just a theory of alternative possibilities and was therefore incomplete. Besides, Einstein disliked the new notion of quantum entanglement allowed by the uncertainties of the clouds of probability wave function.

ENTANGLEMENT

The mysterious phenomenon of entanglement is a quantum state of connectivity through all of eternity between two previously entangled particles that are now separated in space and/or time. When one of the particles is observed and its wave function collapses, an instantaneous and corresponding effect occurs to the other that happens faster than the speed of light. If an electron is observed to have a spin up, then the other electron absolutely has to correspondingly have a spin down.

Both members of the entangled pair immediately know exactly what is happening to the other and behave accordingly, although they are separated and have no way of transmitting information about their experiences to each other. To Einstein's mind, the probability wave function could not possibly act at two different places and not violate general relativity, which requires everything to travel no faster than the speed of light. He resisted the idea.

Einstein, joined by the physicists Boris Podolsky and Nathan Rosen, responded to the Copenhagen interpretation with another thought experiment, known as the EPR paper after its authors' names. They argued that a hidden variable or factor must be at work, just like what prevails in classical mechanics. To them, a preexisting element of reality had to exist to explain the "spooky relatedness" of particles that know immediately what happens to one another without an identifiable means of sharing information.

In 1935, when John Stewart Bell (1928–90), a Northern Irish theoretical physicist who did research at CERN, showed the community that the EPR paper could be translated into practical laboratory experiments, physicists got busy. Bell's theorem states that "if a hidden variable theory is local it will not agree with quantum mechanics and if it agrees with quantum mechanics it will not be local."

To make a long story short, all predictions offered by quantum mechanics have never been proven wrong by any experiment based on Bell's theorem, and so far, no local variables or local hidden elements of reality have been found to explain the "spooky" behavior of entangled particles. The Bell theorem tolled the demise of EPR and Einstein's relativity rule that nothing can travel faster than light. The Copenhagen interpretation survived, and Bohr became the champion of the quantum movement, which speedily marched on, thriving and profusely blossoming, not only in physics laboratories of academia, but in research laboratories of industry and commerce as well.

FIELDS OF PARTICLES

To accept particles as the theoretical building blocks of material reality required the acceptance of quantum forces to implement their interaction. The concept of force fields was born. The idea was considered necessary for the creation of particles, with each field associated with a certain particle and vice versa. The scheme presupposes an all-pervading and unchanging backdrop that allows "jittery" ripples of energy to crystallize into materiality in response to conscious observation.

Don Lincoln, a senior particle physicist at the Fermi National Accelerator Laboratory ("Fermilab") near Chicago, who splits his time between Fermilab and CERN doing experimental particle research, posted an article on quantum field theories (QFT) on the website NOVA in 2013. In it, he described a specific particle as simply a localized vibration of its own specific field that spans all of space: the electron in the electron field, the quark in the quark field, the photon in the electromagnetic field, the gluon in the gluon field, the neutrino in the neutrino field, and so on.

More particles and fields are predicted to be found, and it hurts the brain to struggle to envision "sparticles" such as neutralinos, Higgsinos, photinos, winos, zinos, gravitinos, and many others more—each particle in its own specific field.

In principle, the quantum force field is tied to Heisenberg's uncertainty principle. The field's variable energy density is a result of its constant particle and antiparticle production and mutual annihilation and dissolution. The energy fluctuations seem regular and smooth at low-power magnification, but look uneven when subjected to the LHC, the largest and most powerful microscope on earth. The detection of irregular field-energy jitteriness is akin to looking at a spotless face through cataract-clouded vision, only to find out after successful eye surgery that the beautiful satin-smooth skin is actually riddled with large pores and blemishes.

Energy in this quantum force field materializes into perceived particulate state upon human conscious observation. If energy intensification caused by conscious observation is not enough for the brain to attain conscious perception, then the short-lived particles dissolve back into quantum probability waves of the immutable and pervading framework of infinite energy—just as a drop of rain loses its individuality upon joining and becoming one with the sea.

The idea of the quantum force field is new and old at the same time. In ancient traditions, it is variously called *chi* or *qi* in China, *ka* in ancient Egypt, "life force" in various cultures, and *prana* in India (*prana* means constant vibrational movement), all of which refer to primary energy or original creative power.

6
The Quantum Enigma

*The world is not as real as we think.... My personal opinion is that the
world is even weirder than what quantum physics tells us.*
—*Anton Zeilinger*

A background history of the nature of light is necessary in our
quest for knowledge about the quantum. Light exists as a particle
(a photon) and as a wave (an electromagnetic wave) at the same
time. Yes, all at the same time, in a dual mode of being. How can
that be? That question has roots that date back to the Greek
philosophers Democritus (c. 460–c. 370 BC) and Aristotle (384–
322 BC), the Roman philosopher Lucretius (99 BC–55 BC), and
the Arab Muslim scientist Alhazen (also spelled Ibn al-Haytham,
AD 965–1040), all of whom presented diametrically conflicting
views on the nature of light.

Aristotle was the first to suggest that light travels as a wave-like
disturbance in air, opposing the view of the first atomist,
Democritus, who thought that all things, including light, were
made of invisible, indivisible particle units called atoms. A few
centuries after Aristotle's introduction of the wave-like
propagation of light, Lucretius proposed that light came from the
sun as solar atoms, just as Democritus had suggested before him.

In the early eleventh century AD, Alhazen wrote the first
comprehensive description of light as rays, a stream of particles
traveling from the emitting or reflecting object to the eye, in his
seven-volume treatise *The Book of Optics*. That objects emit or
radiate their own light or reflect it from another source was the

exact opposite of earlier hypotheses that held that vision was a result of light from the eyes illuminating the object.

Alhazen was a mathematician, physicist, astronomer, and philosopher whose work on optics and the anatomy of the eye and its role in vision led to his renown as the father of optics and ophthalmology. He is widely considered to be the first true scientist to pioneer the scientific method of investigation.

Sir Isaac Newton, a mammoth figure in mathematics and physics in the late seventeenth and early eighteenth centuries, proposed the existence of light as a particle in his book *Opticks*, published in 1704. He called such particles "corpuscles." On careful scrutiny, however, he had a problem. A straight rod that we stick in water looks broken, or refracted, at the point of entry into the water. On the other hand, we also see an unbroken image of the rod on the surface of the water, which is its reflected image. When a ray of light traveling as corpuscles hits the surface of water, some of it gets refracted, and some gets reflected. If light traveled as corpuscles in a straight line, then wouldn't it all get refracted or all get reflected? Light corpuscles should not do both at the same time. What would make a medium treat light in two different ways?

Newton's way of thinking at that time was incredibly predictive of the future concept of the nature of reality—the quantum wave-particle duality. He explained the phenomenon as a function of a gate-like procedure of admitting light in (refraction) at first and then closing the gate on the next light corpuscle (reflection), which he called "alternate fits of transmission and reflection."

Newton studied the effect of prisms on light. With one prism, he split white light into the seven colors of the rainbow and then put them back together again with another prism. He observed that an object's color was dependent on the color of the light shining on it. He decided that color is a property of light and not of the object itself.

He also studied the reflection and refraction of light in a medium. Again, these experiments presented a wave problem. If light traveled as corpuscles, then why, when shone through glass, would some of the light get reflected? Corpuscles traveling in a straight line should either pass through the glass or get stopped. According to Newton, in that instance, light provoked a "wavelike disturbance" in the "ether." He correlated the degree of the ether's disturbance to the color of light. In the red spectrum, the light corpuscles possessed the greatest mass and they made the greatest disturbance. In the violet spectrum, the light corpuscles had the least mass and created the least disturbance.

It is amazing how close he was to figuring out that the rainbow color of light that his prisms separated and the wavelike disturbance that each color stirred in the ether is a function of wavelength. He was on to something, although he did not fully realize it. His explanation was not satisfactory. But because he was Sir Newton—someone who was widely recognized as the greatest mind of his time and an undisputed authority whose word was good enough—not too many had the gall to buck him. And although it was Alhazen who wrote first that light rays were streams of particles, or corpuscles, the corpuscular theory of light stuck with Newton's name through history.

Christiaan Huygens (1629–95), a Dutch physicist, astronomer, mathematician, and contemporary of Newton's, offered an authoritative, albeit lonely, dissenting voice. Considered a leading scientist of his time, Huygens is credited with discovering the Orion Nebula, Titan, and the rings of Saturn; with inventing the pendulum clock, the most accurate timekeeper from 1656 until 1930; and with founding the wave theory of light. In 1678, he proposed that light travels as waves in ether, just as sound waves do in air, which is why, just as in the case of sound, no evidence has ever been found of collisions of light, which would surely happen if light traveled as corpuscles. Even so, he did not publicly refute Newton's light corpuscles, which dominated the scene.

Huygens's work in this field lay buried in obscurity and was almost forgotten for many years.

Decades later, a young French physicist and engineer named Augustin-Jean Fresnel (1788–1827) worked on Huygens's wave hypothesis and presented it with a more refined mathematical expression in which he made clear that the refraction of light depended on the property of the material it was traveling through, its speed changing with variations in the densities of the medium. For instance, light travels faster in air than in water because these media have different densities and different abilities to bend light; now referred to as the "refractive index" of a specific material. Still, the Huygens hypothesis of light as a wave stayed hidden under the long shadow cast by Sir Newton's eminence, and the idea had to wait for Thomas Young (1773–1829), another Englishman, to enter the scene.

An unbelievably bright physician and polymath who was fluent in more than a dozen languages, Young pursued his diverse interests unhindered, thanks to the financial freedom afforded him by a wealthy uncle. Like Huygens, Young also believed that light was a wave. In 1803, he undertook experiments to prove it. The laboratory result—a double-slit light-interference pattern—brought him international fame and authority as a scientist that continues to this day.

In his classic research, light from one source was shone on an opaque screen with two narrow slits positioned close together. These slits let light through and became two separate sources of light, both traveling concentrically as waves and meeting each other along their way to a second opaque screen. At points where their crests arrived at the same time, a summation of their brightness happened, just as the waves in water get higher where two wave crests meet. At points where a crest from one slit arrived at the same time as a trough from the other slit, they canceled each other out. On the second screen, the effect was that of alternating light and dark bands, respectively showing the effects of the summation and cancellation of the light waves (figure 5). Light corpuscles do not and cannot produce an interference pattern. Just like in the case of real-life corpuscles such as red blood corpuscles, if they hit one another, they bounce off one another. They do not result in summation or subtraction effects; only waves do that.

Young's double-slit light experiment sealed the fate of Newton's theory. The shadow cast by his corpuscles gradually faded away in the light of Young's interfering waves. There was no more doubt that light travels as waves.

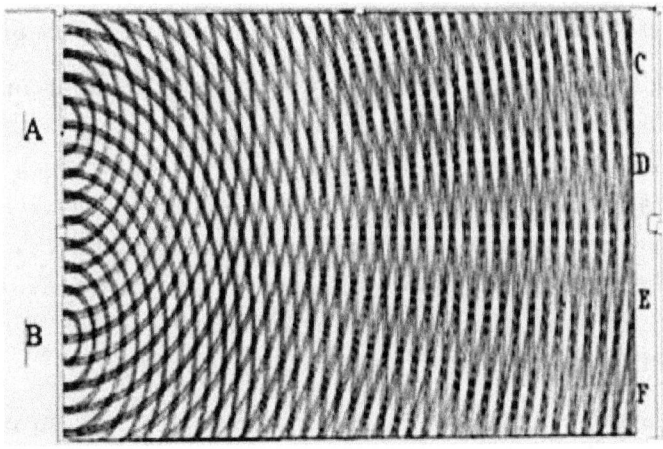

Fig. 5. Thomas Young's double-slit light-interference pattern, presented to the Royal Society in 1803. Light from one source enters at slit A and slit B as waves and spreads from each slit and interferes with the other to create light (C and D) and dark bands (E and F). The light and dark bands alternate on the screen. Source: Thomas Young.

The reaction to Young's double-slit experiment was ferocious. How dare he challenge Newton's corpuscles? However, other scientists subsequently confirmed his findings, and the wave property of light eventually gained acceptance even among the most loyal Newtonian believers and fans in England and beyond.

The nature of light got into troubled territory again when Einstein further studied Planck's blackbody oscillators, which quantized the emission of light energy. In December 1900, the forty-two-year-old Planck solved the blackbody problem with his $E = h\nu$ equation. Just a short time thereafter, Einstein announced his discovery of the quantization of none other than light itself. After Young's double-slit light interference experiment eclipsed Newton's light corpuscles, Einstein's quantized photons traveling as particles came bouncing along to take center stage. This created a perception problem. Light had gone back to being identified as a particle, despite having been established as a wave. Scientists were bothered and bewildered. Is it a wave? Is it a particle? Is it both?

De Broglie believed that light was both. Working on Bohr's quantized electron orbits in 1923, he assigned waves with specific frequencies (and thus wavelengths) to the orbits, creating a model of an atom in which the electron existed as a standing wave around the nucleus. With this 1929 Nobel Prize–winning doctoral thesis on the wave-like behavior of matter, the thirty-seven-year-old prince accomplished the insertion of the enigma of quantum wave-particle duality into people's consciousness.

The wave nature of the electron was proven in an electron-radiation diffraction experiment in 1927 and definitely shown in a double-slit experiment using electrons instead of light in 1961.

In 2015, Fabrizio Carbone and his colleagues at Switzerland's École polytechnique fédérale de Lausanne used a composite of a series of images at variable time delays to capture the behavior of light as a standing wave and as a particle. Although what they found was not an actual image of a single photon caught red-handed acting as a particle and as a wave at the same time in sure violation of Heisenberg's uncertainty principle, it decidedly demonstrated the wave-particle duality of light.

De Broglie's thesis on the dual existence of the electron as a particle and a wave facilitated Schrödinger's development of the quantum wave function equation, which presents matter of any size moving at any speed that would allow the wave with the highest-probability amplitude or energy density to materialize as a particle, at which point all the rest of the probability waves would simultaneously collapse to zero.

A CAT IN A BOX

As discussed in chapter 5, Schrödinger thought that the existence of the electron in a wave and particle "superposed" state could extend to atoms and everything else in nature—made of atoms as they are. But then, following logic, this idea could essentially deny reality. Objects would all be reduced to waves of probability until they were observed, which to him, at that time, was a ludicrous idea. (Later on, when we talk about consciousness, we will discuss why this particular notion of our reality may be legitimate and is not absurd after all.)

He designed a thought experiment on a large-scale basis, a classical-physics equivalent to the quantum spread-out electron, to point out the idea's ridiculousness: a cat inside a box existing in superposed dead-or-alive probability states. This imaginary cat and box, and in fact the whole contraption itself, exists in a quantum state of probability waves. It contains a miniscule amount of a certain radioactive substance (also existing as waves of probability) with a decay rate of one atom per hour. The radioactive material, had a fifty-fifty chance of decaying or not decaying, and if it did decay, then the radiation releases cyanide, a poison, thus killing the cat. If, after one hour, no decay has taken place, then the cat stays alive. The poor experimental kitty remains in limbo, with equal probabilities of being either dead or alive. Because Schrödinger further supposed that the uncertainty of the quantum could be extended into the macroscopic, and could be resolved by direct observation, an act of observation, like opening the box and looking, would take the kitty's superposed deadness or aliveness out of its fuzzy existence and bring it into the classical world's either definitely dead or definitely alive particulate state.

While it is difficult to make a mental image of the cat's dead-or-alive predicament, the Schrödinger thought experiment did become the mascot for the equations of the quantum wave function of the whole material universe. The theory advanced successfully despite its author's apprehension about the can of worms his wave function had opened. In fact, his extensively used equations propelled the forward course of quantum mechanics.

Still, the all-spread-out electron was difficult for many of his contemporaries to envision. It just did not lend a good pictorial view to the imagination. Enter Max Born to the rescue with a welcome interpretation, which, together with contributions from Bohr and other colleagues, cleared the muddle of an electron smeared all over the place and the sickening image of a cat in a mixed-up state of being alive or dead in superposition with each other. Born explained that the electron wave function is a "state of probability of finding" the electron—a purely statistical affair.

The electron wave has to be approached in terms of multiple probabilities. The chance of finding the electron somewhere is greatest wherever the probability wave amplitude or the energy concentration is highest, and the chance is the least where energy is the lowest. In other words, the appearance of the electron as observed matter in a specific place at a specific time depends on that particular wave of probability having the greatest energy and highest statistical chance among all other waves of probability.

For non-physicists, it is important to note that the probability wave amplitude that scientists are talking about here, though it deals with waves, is a complex computed statistical number, not the height of a wave that we are used to envisioning. Besides, the statistical probability it describes is one in continuous flux, not a definite fifty-fifty chance of being either head or tail in an honest coin flip. The quantum has its own special turf, something quite different from ours.

Born believed that the wave nature of matter, as displayed in the behavior of the electron, implies that the description of matter and of the entire observable universe needs to be in accord with the probabilistic nature of matter. It follows that all material objects exist only as superpositions of multiple quantum waves of probabilities of being until they are observed, at which point they become real to observers in their macroscopic classical Newtonian material world. This oddity later came to be recognized as the observer effect.

Born's new idea was monumental. It meant that the then-accepted classical paradigm about the relationship of celestial objects like the sun and the moon as fixed would have to give way to the enigmatic uncertainty of probability waves. The evolution of life and the cosmos and our picture of the whole shebang are therefore subject to probabilistic uncertainties! The concept was a tough one for scientists to absorb intellectually.

In the chapter on consciousness, we will discuss the mechanism by which the conscious mind can affect these indeterminate affairs and make them materially definite and real, making good what has been consistently taught to us through the ages. According to Matthew 21:22 in the Holy Bible, King James Version (KJV), "And all things, whatsoever ye shall ask in prayer, believing, ye shall receive." And according to *The Kybalion*, a book on Hermetic philosophy published in 1908 by the Yogi Publication Society in Chicago, Hermes said: "You are dwelling in the Infinite Mind of The All, and your possibilities and opportunities are infinite, both in time and space."

With today's electronic technology, the genuine uncertainty of the superposition state in which an electron wave function is truly spread out explains the phenomenon called quantum electron tunneling, in which an electron crosses a mechanical barrier without actually jumping it or digging through it. This is the principle behind transistor systems which make all electronic devices work: turning them on or off, changing the intensity of light or sound, showing images and movement, or allowing memory storage or calculation features, to name a few functions.

The quantum mystery deepened and had physicists and other scientists unsettled. Einstein was unhappy. He felt that there had to lurk a hidden key and fundamental mechanism underlying the whole chancy game of probability waves, and it was only the limitation imposed by the incompleteness of quantum theory and the weakness of human understanding that allowed this pesky quantum uncertainty to impose its rule on fundamental physics. He was moved to utter one of his many quotable statements: "God does not play dice."

Likely unbeknownst to many early-twentieth-century quantum physicists, Schrödinger's probability wave equation truly allows the wave function to exert an effect faster than the speed of light and to offer a clear explanation for the enigma of quantum entanglement. "Spooky action at a distance" was Einstein's derisive remark when he was presented with the thought experiment on quantum entanglement, which is now considered an intrinsic principle of quantum physics. The phenomenon clearly disregards the classical laws of physics which govern ordinary macroscopic events. If accepted as law, however, then the idea can help people understand how the observer effect brings to pass their different personal realities.

The Schrödinger probability wave function equation went on to be successfully implemented in science and industry and has been confirmed with precision in the microscopic world of electrons, atoms, and molecules. Outside the world of the tiny, though, material objects appeared to behave strictly according to the laws of classical Newtonian mechanics—that is until 2010, when researchers induced a tiny metallic semiconductor paddle, visible to the naked eye, into a quantum superposition state of vibrating and not vibrating at the same time: a strange property of matter compared to its normal everyday-life appearance.

This breakthrough clearly indicates that the rules of quantum mechanics bear upon classical Newtonian mechanics as well, and it enhances our grasp of the hidden dynamics by which natural physical laws are exposed in our own day to day reality. The idea foretells practical applications ranging from faster-than-light information exchange, quantum memory storage, and various tools for further inquiry into the root of the quantum enigma.

Recently headlined in physics news is hydrogen, the simplest atom, which has only one proton and one electron. The reason for the excitement is the image shown in figure 6, which displays the hydrogen electron and its probability waves or wave function outlined as alternating bright and dark rings of interference pattern. Accomplishing the task was not a simple process of taking a shot with one click of the camera. One particular quantum property of the electron, the Heisenberg uncertainty, totally prevents it from being directly observed by an instrument. Even if one could take a picture of the atom (a wishful thought) and could see the electron in a specific position in its orbital, the single image would not provide us any sense of its wave function.

Aneta Stodolna, a Polish physicist, and her fellow researchers at the FOM Institute for Atomic and Molecular Physics (AMOLF) in the Netherlands cleverly got around these limitations and obtained an electron probability wave image by simultaneously capturing countless imprints of electrons released from hydrogen atoms by using a scientifically and technologically sophisticated procedure. According to her senior co-researcher, Professor Marc Vrakking (b. 1963), director at the Max Born Institute for Nonlinear Optics and Short Pulse:

Since isolated hydrogen atoms do not naturally exist here on earth, they were first produced from hydrogen sulfide, or the H_2S molecule, by photodissociating [disengaging molecules using light] these hydrogen sulfide molecules with ultraviolet light. Then, using laser beams from two sources precisely tuned to the resonant frequency of the hydrogen atom, the electron was excited (photoexcitation) and then cleaved from the atomic nucleus (photoionization), resulting in a hydrogen ion (only the single-proton nucleus remains) and a free hydrogen electron.

The hydrogen electron the research team produced went on a path toward the two-dimensional detector of the quantum microscope, located roughly a foot and a half away. Hitting the detector effectively led to an "observation" of the electron, and that encounter, as discussed earlier, implemented the collapse of the wave function of the electron and forced it to appear as a particle in a specific position, in accordance with the statistical probability of its appearance.

The images the team obtained from numerous electrons in their most probable orbital positions were incorporated together, and out came an eerily beautiful snapshot-like picture of the hydrogen atom (sans nucleus, though one can go ahead and imagine it in the center) for the first time ever! The image vividly shows dark rings where the probability of finding the electron is zero and bright rings where electrons have the highest probability of appearing.

The composite picture of countless electron wave functions—collapsed in their most probable orbital positions—provides, in toto, a clear and elegant visual representation of the hydrogen electron's hazy cloud of probabilities and wave function, allowing this distant and nebulous theoretical concept to engage us, no matter how contrived, with its palpable and friendly presence.

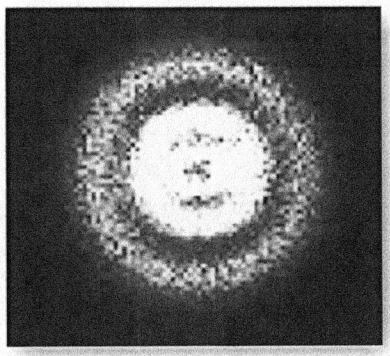

Fig. 6. A direct macroscopic image of the hydrogen electron's quantum state. The interference pattern of the wave function shows dark rings where the electron is least likely to be found and light rings where it has the greatest chance of appearing. Image reprinted with permission from Stodolna et al., "Direct Observation of the Nodal Structure of Stark States." Physical Review Letters, *vol. 110 (2013). Source: American Physical Society / Aneta Stodolna and Marc Vrakking.*

Every bit of this astounding accomplishment was made possible because the natural vibration of the hydrogen atom had been identified. Today, scientists have laser-emitting machines that can be tuned to duplicate this frequency in order to excite the electron and boost its energy enough for it to escape the tight hold of nuclear attraction. This feat has made Einstein's "photoelectric effect" and "stimulated emission" come to life. It is nothing short of amazing how much the growth of knowledge and high-tech gadgetry has empowered humanity in less than a century.

THE OBSERVER EFFECT

Around 1934, Pascual Jordan, the physics whiz kid introduced earlier, said of the mysterious phenomenon known as the observer effect: "observations not only disturb what is to be measured, they produce it." That was a radical statement then, but now, the idea has already been extended into research on the very nature of consciousness.

Quantum physics tells us that photons or electrons will go through double slits to hit a screen as waves when unobserved, yet they will go through as particles and be recorded on the screen as dots in the presence of an observer. In a 1998 issue of Nature, the physicist Eyal Buks and his colleagues at the Weizmann Institute of Science in Israel reported an experiment on multiple electrons sent through a double-path interferometer (a light-interference measuring device) with a "which-path" detector set up near one of the paths. What they found was that by dialing the strength of the electrical supply up or down to vary the sensitivity of the detector, they could also vary the visibility of the wave interference pattern. When they ratcheted the intensity of the observation upward by increasing the sensitivity of the instrument, the interference pattern became fainter, and when they lowered the intensity of observation, the visibility of the pattern was clearer, thus explicitly demonstrating the observer effect—in varying degrees, to boot—before a critical increase in the instrument's sensitivity forced electron particle materialization.

The intensity of observation in the laboratory decidedly affects the outcome just as the intensity of desire, intention, and focused action greatly affects our experienced reality.

Early in the history of the quantum, observation was said to be accomplished when a microscopic entity came in contact with something macroscopic, be it a piece of equipment or a conscious being. Quantum entanglement dictates that any such contact implicates all components of the observed and the observer—the measuring apparatus and the conscious mind included—and incorporates them all into a single wave function. In 1932, the Hungarian-American physicist, mathematician, computer scientist and polymath John von Neumann (1903–57) proposed that the combined wave function of the quantum entangled object and the measuring instrument collapses into a definite state of being upon measurement in the presence of a conscious observer.

Three decades later, another Hungarian-American, physicist Eugene Wigner (1902–95), offered that consciousness is the final observer that engenders the collapse of the wave function. Neither von Neumann nor Wigner indicated how this collapse might come about. Moreover, in order to bypass the question of consciousness in the lower forms of life, scientists and other thinkers have narrowed this broad territory of observer-induced collapse of the wave function to involve human consciousness alone as the final arbiter, a mode of thought that allows the superposition state to involve the whole measurement apparatus, the human organism and human consciousness included. Thus, as far as our discussion of reality is concerned, human consciousness, as an entity distinct from the material human body, is, for all practical purposes, the final measuring device.

How does this observation of the tiny quantum in physics laboratories translate to our world of sticks and stones? Most physicists shy away from integrating the laws of quantum physics with those of our day-to-day experience, never mind the stress arising from unquenched curiosity. Without a doubt, the situation is disconcerting to many, and increasingly so. According to Rosenblum and Kuttner, the physicists mentioned earlier, "some of us, as physicists, or just as wonderers, ponder the meaning and try to understand what is really going on. This has long been an attitude of many eminent physicists.... It's an attitude that today [is gaining] acceptance."

And why shouldn't this mystery pester a physicist's waking thoughts? We are told that everything in this universe is made of quantum energy jitters, frequencies which make up subatomic particles, which make up atoms, which make up molecules, which make up tissues, which make up organs, which make up systems, which make up organisms, including the human organism. The rules of quantum mechanics should apply in a very basic sense to the macroscopic domain just as they do to the microscopic.

To physicists, the gaping divide between theoretical constructs and laboratory observations of the very tiny and the day-to-day experience of macroscopic materiality must feel intuitively unacceptable and must invariably haunt their thoughts.

Certainly, we have to respect the scientists' reluctance to go beyond their field of theoretical and experimental expertise. Be that as it may, in the end, the mysteries remain behind closed doors. This quagmire is a glaring indication that hard science is not and never was in a legitimate position to explain nature fully. Therefore, the world of science cannot rightfully deny outright the existence of personally experienced phenomena that could not be proven or duplicated in scientific laboratories of gadgets and chemicals. The skeletons in physicists' intellectual closets have to inevitably rattle too loudly to be summarily ignored.

On the other hand, the mysteries of the quantum have always been in full experiential view of mystics, master believers, and transcendental meditators. Maybe, in the future, as the discomfort grows with having to constantly bang the closet doors on skeletons that stubbornly keep on sticking out, scientists will be emboldened to investigate the dimension that transcends material nature, as some colleagues before them have already dared to do.

THE QUANTUM WORLD AND EVERYDAY REALITY

Quantum mechanics has found definite applications in daily life. They include simple electric light switches, transistors, lasers, flash drives, computer microchips, mobile devices, household cleaning robots, artificial intelligence (AI), GPS, Bluetooth, Wi-

Fi, quantum chemistry, electron microscopy, magnetic resonance imaging (MRI), nuclear magnetic resonance (NMR) spectroscopy in molecular physics and crystalline and non-crystalline material research, superfluidity, computational chemistry, biophotonics, and quantum cryptography; not to forget nuclear reactors and bombs. The basic principle underlying all of these modern-day applications is quantum mechanics.

The early implementation of quantum computations by high-tech industry enormously incited many advancements in science. The application of the quantum mechanics eventually extended to commerce and brought an unprecedented deep understanding and expert utilization of biological and chemical quantum functions such as photosynthesis, vision, nuclear medicine, DNA function, genetic engineering, designer pharmaceutical drugs, removal of toxic gases such as carbon monoxide, and many more.

As of this writing, rapid computation is still being performed through the binary digital method, which encodes electrical signals converted to series of the digits 1 and 0 in order to transfer and store information. Research on quantum entanglement at ambient room temperature promises ultrafast computing, processing, and storage of quantum information in the very near future.

Quantum theory is the most accurate among all theories that physical science has ever advanced, each of its predictions having been confirmed to the highest degrees of accuracy in experiment after experiment. And what a life-enhancing revolution the theory has brought us. Let us take stock of how it has affected our lives.

Close your eyes and imagine this: You are quietly standing in line at an electronics store checkout counter with LED light bulbs in your basket. In front of you, a young woman in an executive outfit has a thin laptop computer and a laser pointer, which the clerk is scanning. "She must be giving a PowerPoint presentation," you muse distractedly as a familiar movement of Chopin's

Nocturne Op. 9 No. 2 wafts faintly across the store, broadcast wirelessly through Bluetooth. Now open your eyes.

The scene you have just pictured in your mind's eye is so everyday humdrum, mundane, and absentmindedly automatic to the point of being unconscious that you do not give it any thought at all. That is, until you realize that your imaginary scene would not have been possible had Sir J. J. Thomson not discovered his plum-in-a-pudding electron.

The fast age of electronics dawned with his totally life-changing discovery. Now, people are enjoying a lifestyle that J. J. Thomson most probably had no inkling was coming. In fact, everything that characterizes the electronic age of today proceeded from the electron revolution he fathered.

THE WORLD WIDE WEB

The World Wide Web (not to be confused with the internet, an earlier invention[7]) is so familiar that we are often lulled into

---

[7] In the late 1960s, the US Department of Defense funded the Advanced Research Projects Agency Network (ARPANET) to establish a single network of communication between multiple ARPANET research computers in different universities and institutions all over the country. The first computer-to-computer (node-to-node) message was sent in 1969 from the University of California, Los Angeles, research laboratory to the Stanford Research Institute in Stanford, California. This ability to send messages resulted in the development of email and the internet backbone protocol, the "Transport Control Protocol / Internet Protocol" (TCP/IP). The TCP breaks messages into "packets" and IP addresses and then forwards them. The TCP/IP allowed the evolution of a dynamic network of networks that enabled different networks (instead of only node to node within one network) to communicate. The term "internet" is short for "internetwork." With the privatization of the ARPANET and the release of the World Wide Web to the public domain in the early 1990s, the internet rapidly became part of the daily lives of roughly half the world's population.

thinking it has existed for as long as every living person we know has been around. That is not so. It is a recently coined term, born out of necessity. In 1989, a twenty-four-year-old computer-software engineer at CERN named Timothy (Tim) Berners-Lee saw a need for an efficient and quick way by which CERN's collaborating scientists all around the globe could share their research data. Their laboratories were equipped with a host of computers that had no technical capability of communication with one another.

He proposed a project to his division management to develop his idea of a "system for information distribution." Rejected at first, the young computer wonder, boosted by his supportive collaborator, Robert Cailliau, a Belgian computer scientist, persisted. He was finally given the green light in 1990. Not too long afterward, the terms URL (uniform resource locator), HTTP (hypertext transfer protocol), and HTML (hypertext markup language) came into being, and the World Wide Web was born.

From the very beginning, Berners-Lee, who had a far-ranging vision for the application of his information distribution system, aimed at developing a medium that could be used by all people, including those outside the scientific community. It was a noble goal, and in April 1993, CERN announced the release of the World Wide Web to the public domain, making the new and still unknown software available for use by anybody and everybody, thus opening the "www" floodgates to businesses and to the general public—computer geeks and non-geeks alike.

The invention changed forever the way we conduct our very business of living. It thrilled many and frustrated a few like me, who many times would have loved to smash the computer to smithereens on the floor—were it not for the sobering grim recollection of the sticker price. In 2004, Queen Elizabeth II knighted the foresighted father of the World Wide Web for his pioneering work.

The World Wide Web has evolved in many ways since then as more and more people have applied it to their circumstances and whims, constructively or destructively, as exemplified by the many positive as well as negative influences of different social media. Today, just a little more than a quarter century later, the World Wide Web has become inextricably, whether subtly or grossly, woven into the canvas of our reality and is now entrenched as an integral part of our daily lives, thanks to Sir Timothy Berners-Lee.

There is no turning back.

7
Sound, Number, and Geometry

Philosophy is written in this grand book, the universe.... It is written in the language of mathematics, and its characters are triangles, circles, and other geometric figures.
—Galileo

Sound is a mechanical disturbance from the state of rest of a medium, such as air. In the human language, a word is formed by combining sounds in order to communicate an idea. In fact, words in many different languages across many continents that exhibit similarity in meaning have the same root. *Logos*, the Greek term for sound, word, and reason or logic, is derived from the root *leg*, from which came the Latin *lux* (light) or *lex* (law), and the Greek *lychne* (light or lamp). This gave rise to *langi* in Polynesia and *langit* in Malaysia, Indonesia, and the Philippines, their meanings ranging from light to sky and the heavens. In ancient Greek, *logos* meant word or reason, which overlaps with *ratio*, the Latin term for reason, judgment, numbering, calculation, and the relationship between two numbers. The *Oxford Dictionary* gives *logos* two meanings: in theology, *logos* is the word of God or the principle of divine reason and creative order, identified in the gospel of John with the second person of the Trinity, incarnate in Jesus Christ; in Jungian psychology, *logos* is the principle of reason and judgment.

The very first verse of the first chapter of the Sefer Yetzirah, or Book of Creation (the earliest of surviving books on Jewish esotericism), proclaims that the sublime, ever-living God of Israel created the universe by the three *sefarim*, or books: speech, writing, and number. Speech, meaning sound, represents the infinite and unlimited. Writing represents the limited geometry of form and structure; number signifies mathematics as it applies to the harmonic structural arrangement of matter. The Book of Creation contains the ancient Hebraic teachings of the kabbalah, which are grounded on Mosaic temple sacred geometry and Hellenistic Pythagorean sacred science.[8] The kabbalah, in olden times, was an exclusively oral tradition. The term, literally meaning "tradition," is a hallowed compilation of Hebraic esoteric wisdom meant to reveal the relationship between the unchanging, mysterious, and infinite creator, *Eyn Sof*, and His mortal creation: the material universe, humans in particular.

SOUNDING MATTER INTO SHAPES

In kabbalistic cosmology, matter comes into existence by the harmonic geometric arrangement of vibration. The frequencies of

---

[8] Sacred geometry, or sacred science, is a belief that the emergence of the universe from the unknowable and unfathomable void, before time and space, was a geometrical act of God. Sacred meanings are attached to certain geometric shapes, proportions, and symbols, including letters of the alphabet. Sacred geometry was handed down through several generations from Abraham (who supposedly learned from Hermes) to Moses, the leader of the Israelites' exodus out of Egypt. The Hebraic descendants of Abraham have followed sacred science for many years, as exemplified by the specifications in the construction of Solomon's temple and by Hebraic symbols such as the Star of David. Ancient Egyptian sacred science also thrived in the culture of the Egyptian Hebraic Hellenes (descendants of Hellen, the ruler of ancient Thessally in northern Greece), who settled in the different countries ruled by the Greeks, including Egypt. (The ancient word Hellas for Greece was derived from the word *hellen*.) Sacred science was a belief, and a way of life as well, of the Pythagoreans, whose life philosophy was influenced by Hermetic teachings. Pythagoreanism saw a revival during the Hellenistic period (336–30 BC), which started with Alexander the Great (356–23 BC), until the death of Cleopatra in 30 BC, during which cultures of the natives of conquered territories and settlers from Greece and from Asia Minor, southern Italy (Hellenia), Egypt, Mesopotamia, and Persia—blended.

quarks, protons, neutrons, electrons, atoms, and molecules are mathematically governed to establish harmonic and resonant relationships in their geometric crystallization into matter. As well, the spatial and temporal relationships of formed matter are harmonically guided geometric structures.

Sound, number, and geometry figured profoundly in the German polymath Johannes Kepler's *Harmonices Mundi* when he introduced his "polyhedral platonic solids" concept of planetary motion and positions in space. The platonic solids—tetrahedron, hexahedron, octahedron, icosahedron, and dodecahedron—are named after Plato, a Greek philosopher who lived between approximately 424–348 BC and whose teachings laid the foundation of Western philosophy.

To Kepler (1571–1630), the movement of the sun, moon, and planets plotted these geometrical forms in a musical structure. Their relationships with one another had harmonic intervals dictating the sizes of their orbits and their distances from the sun. He related them to power from the sun, which decreased with the planets' distance away from it.

A sphere is the simplest, most perfectly round geometrical form, and like its two-dimensional expression, the circle, it has only one line symmetrically surrounding a central point. One can draw any number of straight lines of the same length from the center, and they will end up exactly on a certain point along the spherical line, neither coming short of nor overshooting it. The sphere is at the same time the most complex shape. Any geometrical shape of any size can be drawn inside a sphere. If one uses the length of the radius to map out points along the circumference of a circle, one will come up with exactly six equidistant points. By connecting every other point of the six, one gets two equilateral triangles intersecting each other, forming a hexagram, popularly called the Star of David. In three-dimensional view, a fourth point extended from a two-dimensional equilateral triangle to create three other equilateral triangles forms a tetrahedron, the basic and most structurally stable geometric form, which requires the least energy to maintain stability.

Outward extensions from the tetrahedron result in the *stella octangula* (or star octahedron), then the dodecahedron, and then the icosahedron. The points of a tetrahedral platonic solid circumscribe a sphere, and the inner faces of the center define another, smaller sphere. As a result, another small tetrahedron can be nested inside the inner sphere, and then a smaller sphere inside the small tetrahedron, and so on. This nesting concept exists in different cultures. When twenty-first century physicists studied the equations of James Clerk Maxwell, they discovered a tetrahedron inside a sphere in the three-dimensional internal structure of the electromagnetic wave.

Five innate vibrational frequencies have been identified in the sphere. From each frequency originates a platonic solid according to resonance, thus orchestrating the geometric crystallization of matter. This process results in stable and functional structures, whether we are talking about the simplest elements or rocks, plants, or the human body. The architecture of all of nature is geometrically framed through resonant vibration.

Vibration is a periodic motion or an oscillatory displacement of a medium in alternately opposite directions from the point of equilibrium when that equilibrium has been disturbed. The vibration's frequency is the number of oscillations per second, usually expressed in hertz (1 Hz = 1 oscillation per second). Its wavelength is the distance between two successive crests or troughs, typically expressed in meters. The amplitude of a wave is the maximum deviation of a point in the wave from the zero or rest position (figure 7).

A wave is a periodic motion along the same path, each displacement cycle taking exactly the same time from one rest position to the next, transporting energy as it travels from one place to another. A transverse wave has oscillations at right angles to its direction of propagation, as happens in the up-and-down movement of a rope being wiggled vertically. Light is a transverse wave in the electromagnetic field and is propagated as quanta of light, or photons. The brightness of light depends on the height or amplitude of the electromagnetic wave, which reflects the number of photons per amplitude unit area. The color of light depends on the wave frequency. The higher the frequency (i.e., with a shorter wavelength), the bluer the light is and the greater the energy it carries per photon.

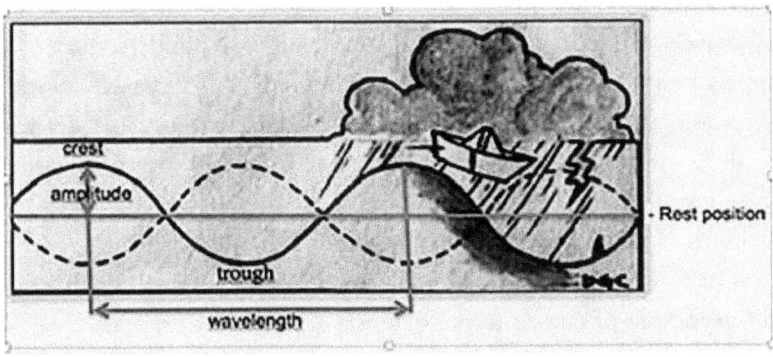

Fig. 7. An artist's depiction of waves in a stormy sea shows the rest position of energy as a line through the center of the oscillations. Wavelength is shown as the distance between two consecutive wave crests (or troughs). The amplitude is measured from the rest position to the highest point on the crest or to the lowest point of the trough. Source: DGC.

A longitudinal wave is produced when the oscillation spreads along the line of travel, exemplified by a Slinky, a toy made of coiled spring. Sound is a longitudinal wave of compressions and rarefactions of a medium, such as air. The loudness of sound and its energy are directly proportional to the square of the amplitude of the sound wave. Throughout this book, the terms "wavelength" and "frequency" denote the vibratory attributes of both light and sound, perceived and unperceived.

STANDING WAVES

When waves travel through the same medium at the same time and they pass through one another, their individual frequencies and amplitudes remain the same. Meanwhile, the medium they pass through shows a net displacement that is a summation or superposition of their amplitudes and of their frequencies. For example, if a pebble is dropped into a still pond, it creates a series of waves of a specific amplitude, wavelength, and frequency. If, subsequently, a second pebble is dropped and it creates waves of the same amplitude and wavelength as the first so that they travel in the same direction at the same time and they are in phase with one another, then the displacement of the water (the crests and troughs of the waves) is the sum total of the displacement from the first wave and the superposed second wave, in accordance with the principle of constructive or destructive interference.

If the same waves are traveling out of phase and they arrive at one point so that the crest of one wave arrives at exactly the same time as the trough of the other, then the resultant effect on the medium is zero, since they would cancel each other out in destructive interference (recall Thomas Young's experiment on light-wave interference). The water shows no disturbance. The extent of the effect of constructive or destructive interference depends on the amplitudes of the wave crests and troughs and how much in phase or out of phase they are with one another.

The same phenomenon is at play when a wave created from one end of a swimming pool hits the wall at the opposite end and creates a reflected wave that travels back toward the original source. If the original and the reflected waves (of the same frequency and amplitude) meet somewhere in the middle, then they become one summation wave that appears to be bobbing up and down in the same place and not traveling anywhere: a standing wave. Non-resonant waves go through the standing wave without affecting it. All standing waves display a point of maximum displacement, whether up or down, called the "antinode," and a point of zero displacement, an area that is undisturbed, still, and motionless, called the "node."

HARMONICS

Everything in nature has a fundamental or primary frequency at which it vibrates at its highest amplitude. The primary frequency is also the lowest, and the wavelength the longest. This frequency is the vibrating object's first harmonic. Riding together with the oscillations of the primary frequency are higher resonant frequencies called overtones, sounds which can be harmonic or inharmonic. When oscillations of two slightly different recurring frequencies regularly interfere with each other, the summation and subtraction of their amplitudes result in the formation of a new slow and regularly recurring waveform of lower frequency and amplitude. This wave is called the beat.

In a vibrating object such as a guitar, the ends of the strings are fixed in place at the bridge and at the nut or at any one fret of the guitar neck. The vibrations, necessarily of the same frequency, are reflected back and forth from the fixed non-vibrating ends, which are the nodes. The distance between the nodes at the two fixed ends (the bridge and nut) is half a wavelength and is the guitar's first harmonic. A full wavelength of vibration requires a third node in between the two, giving the pattern two antinodes: one above the rest position and one below (3:2). This constitutes the second harmonic. The frequencies of the harmonics have a whole-number ratio to one another. Non–whole number frequency multiples are inharmonic. If the primary frequency is 400 Hz, then the second harmonic is 800 Hz (3:2), the third harmonic is 1,200 Hz (4:3), and then the fourth harmonic is 1,600 Hz (5:4).

Harmonic frequencies are consonant—that is, they reach the eardrums at regular intervals, which explains their pleasing harmonious sound. Non-harmonic frequencies are dissonant and irritating to hear. Shown below in figures 8a–8d is an artist's depiction of ocean waves limited at both ends, with the reflected wave shown in dashed lines traveling in a direction opposite to the incident wave. The incident and the reflected waves are of identical frequency and form standing waves. An increase in the frequency (from 8a through 8d) shows the harmonic progression.

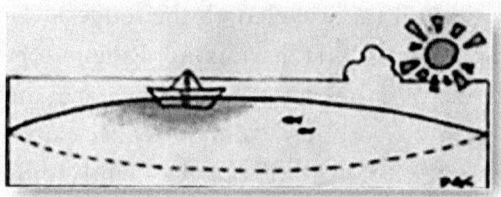

*Fig. 8a. First harmonic: two nodes and one antinode (half a wavelength).
Source, figures 8a–8d: DGC.*

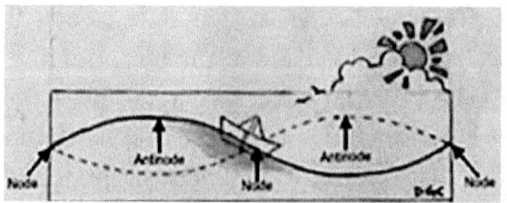

*Fig. 8b. Second harmonic: three nodes and two antinodes (one full
wavelength). The nodes are at the fixed ends of the wave and in the middle at the
position of equilibrium. The antinodes are represented by the wave crests and
troughs. The incident wave (solid line) and the reflected wave (dashed line) are of
the same frequency, traveling in opposite directions and forming standing waves
that oscillate in place.*

Fig. 8c. Third harmonic: four nodes and three antinodes.

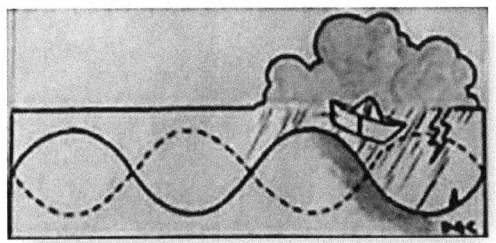

Fig. 8d. Fourth harmonic: five nodes and four antinodes.

CHLADNI FIGURES

That letters, numbers, and spoken words have vibratory hallmarks was provided proof in the twentieth century. Ernst Chladni (1756–1827), a German physicist, musician, and contemporary of Mozart's, discovered that if he strummed the bow of a violin against the edge of a metal plate until a resonant vibration was achieved, the whole plate would vibrate except on certain lines and spots of standing waves or nodes. Thinly spread sand on the plate would move away from the vibrating parts to settle on the still non-vibrating nodes, creating areas of empty spots and sand collections that exhibited standing wave patterns that were specific to the frequency of vibration produced on the metal plate (figures 9 and 10). Chladni's 1787 book, *Discoveries Concerning the Theory of*

Music, laid the foundation for acoustics, a category of science on the properties of sound.

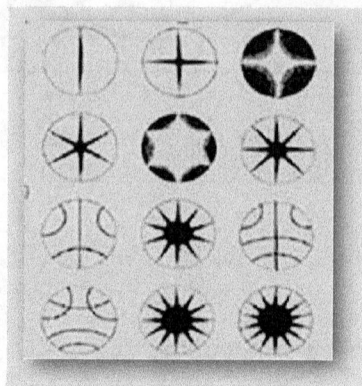

Fig. 9. Image from Ernst Chladni's Die Akustik *(1802). Chladni figures are patterns formed by fine powder on vibrating round plates.*

Fig. 10. Image from William Henry Stone's Elementary Lessons on Sound *(1879). The image shows Chladni figures of sand collecting in nodal areas of a standing wave on a metal plate set to vibrate by strumming a violin bow on the plate's edge.*

Hans Jenny (1904–72), a Swiss physician and mathematician, re-created Chladni figures from fluids, powders, and pastes using crystal oscillators and the human voice. He also used spoken words or musical vocal intonations via a "tonoscope" (a cardboard pipe and crystal oscillators) to set his plates and membranes vibrating. He demonstrated that when the alphabets of ancient languages, such as Hebrew or Sanskrit, were spoken, the sand formed patterns of their actual written symbols. To him, the pattern formed by sound vibration "is not an unregulated chaos." Rather, it "is a dynamic but ordered pattern."

In a two-volume book titled *Cymatics: The Study of Wave Phenomena*, Jenny wrote, "The more one studies these things, the more one realizes that sound is the creative principle. It must be regarded as primordial. No single phenomenal category can be claimed as the aboriginal principle. We cannot say, in the beginning was numbers or in the beginning was symmetry, etc., etc. They are not themselves the creative power. This power is inherent in tone, in sound." Jenny's vast collection of images has been most beautifully displayed and extended by the German photographer Alexander Lauterwasser in his visually impressive 2011 book, *Water Sound Images*.

THE SOUND OF THE SPHERICAL BIG BANG

In *The Secret Doctrine of the Kabbalah*, Leonora Leet presents the kabbalistic description of the very beginning of creation:

> Know that before the emanated things were emanated and before the created things were created, the pure, divine light filled all of existence.... Then *Eyn Sof* contracted himself into a central point with His light in the middle and then removed Himself to the sides encircling the point at the center.... This contraction, equidistant all around the point at the center, formed a void in such a way that the vacuum was spherical on all sides in equal measure.

The parallelism between ancient kabbalistic cosmogenesis and modern cosmology is obvious. One may compare the central contracted point of light inside a sphere with the infinite energy density of the core of the vortex: the singularity that birthed the universe in a big bang. One can also liken the removal of *Eyn Sof* to the side to form a spherical void with the rapid inflationary expansion of the cosmos in the first moments of its birth.

In the Holy Bible, according to John 1 verses 1–3 (KJV), "In the beginning was the Word and the Word was with God and the Word was God. He was in the beginning with God. All things were made through Him and without Him was not any thing made that was made." The Holy Bible says this in Genesis 1:3 (KJV): "And God said, let there be light: and there was light."

The beginning was itself vibration (of the spoken word: "Let there be light"). The spherical void was then filled with light with the very first spoken word. The periodicity of their vibrations guided the formation of the primordial plasma of ionized gas and the harmonic geometric formation of quarks, protons, neutrons, electrons, and atoms as the plasma cooled. In Eastern belief, cosmogenesis is associated with air, fire, and water, which we can relate to sound, light, and primordial plasma in big bang theory.

Sound is primal. Our voiced expressions of satisfaction such as "aah" need no explanation and do not need to be taught. When my very little grandson let out an "aah" sound after drinking a glass of cold water one hot summer day, it seemed natural. It was primal. The mantra *om* or *aum* is primal sound. Eastern monks and other meditators use this sound to vibrate the whole body and mind in order for the meditator to transcend the conscious mind and connect and be one with universal consciousness.

Language originates from primal sounds, as do the alphabets of ancient languages such as Hebrew, Sanskrit, Aramaic, and other primitive languages. Aramaic is a group of many related languages and dialects descended from the ancient Semitic languages and is the ancestor of the Arabic, Phoenician, and Hebrew languages. The Phoenician script, consisting of twenty-two consonants, is the mother of all the alphabets used in languages such as Arabic, Greek, Slavic (which uses a Greek-derived Cyrillic script), and Latin. The Greeks improved on the Phoenician alphabet by adding vowels, and the Italian natives of Latium (centered on today's Rome) used the Greek and Etruscan scripts to develop the Latin alphabet with twenty-six letters that we use today.

With the growth of the Roman Empire came the introduction of the Latin alphabet all over Europe by the Roman Catholic clerics. By the seventh century AD, the alphabet was already in widespread use. It became more standardized after the printing press (invented in the fifteenth century by Johannes Gutenberg) was brought to Great Britain, followed by the publication of the first English dictionary by Robert Cawdrey in 1604.

Several alphabets and words, especially names, in the ancient languages are considered sacred. In Hebrew, the name of God is expressed in the four Hebrew letters YHWH (Yahweh) or JHVH (Jehovah), known as the tetragrammaton. The name was considered too sacred to be uttered by the early Israelites, so they substituted Adonai or Elohim for it. Adonai denotes Lord, and Elohim, God. In Christian religious spoken and written words, English-speaking believers use Lord, God, or (in Latin) Dominus.

Any pattern formed by sound—such as language, music, geometric figures, and others, natural or otherwise—can be converted into a number-sequence format, a method familiar to those who work in the technology sector. The melodies of music, the exquisite nuances of art, and the delicious flow of words in poetry are recognition of and connection with the representations of universal harmonic number patterns or archetypes in the realm beyond perceived space-time.

All things in the cosmos have a sound and respond to sound. According to NASA, a black hole in the middle of the Perseus cluster of galaxies, 250 million light-years away, sings in B flat, fifty-seven octaves below middle C on the piano (figure 11). The sound is a million billion times lower than the deepest audible sound, the lowest frequency in the universe ever detected. To put this in perspective, if the lowest sound that the human ear can hear is in the frequency of one cycle every twentieth of one second, then the song that Perseus sings is at an unbelievably low frequency of one cycle every 10 million years. Nobody can ever hear sound that low (nobody can stick around that long either).

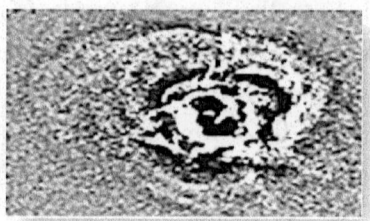

Fig. 11. X-ray image of sound waves emanating from the Perseus galaxy cluster's massive black hole in the cluster's center. The note is in B flat, fifty-seven octaves below middle C on the piano. Source: NASA, Chandra X-Ray Center (CXC), University of Cambridge Institute of Astronomy (IoA), and A. Fabian et al.

The full import of these low frequency waves is still not fully understood, although one thing is known for certain: the energy carried by these sound waves warms the gas clouds of the Perseus cluster in such a way that the formation of new stars is faithfully maintained, thereby preserving the birthing cycle of new galaxies and galaxy clusters.

THE GEOMETRY OF MATTER

Writing, or form, which in the kabbalah refers to geometry, represents the bounded and finite cosmic matter. Anything at all that can be apprehended by the five sensory organs or that can be comprehended by thought is bounded, limited, or finite. The word "geometry" comes from the ancient Greek *geo*, meaning earth, and *metron*, meaning measurement. In this modern age, geometry is defined as a branch of mathematics that deals with the measurement, properties, and relationships of surfaces, points, lines, angles, and solids.

The familiar geometry taught in schools today is based on Euclid's *Elements*, a 300 BC collection of definitions, axioms, and theorems that includes those of Pythagoras, Eudoxus, and Hippocrates of Chios. Euclidean geometry is expressed in the ordered designs of nature that reveal the organization of energy in the geometrical shapes of all living and non-living inhabitants of the universe, such as tetrahedral diamonds, and in the progression of growth of organisms, such as the logarithmic spirals of nautilus shells. Geometry shows up as well in the orderly patterns of the organisms' movements in space and time: for example, in the graceful leaps and pirouettes of a ballet dancer and in the almost-life-like animation of characters in a Disney film.

The two-dimensional version of the icosahedron displays the "Flower of Life" pattern (figure 12), which is found among inscriptions on the walls of the Temple of Osiris in Abydos, Egypt, and in ancient temples and buildings in India, China, and Europe, especially in buildings associated with the history of the Knights Templar. Leonardo da Vinci (1452–1519) calculated ratios from this pattern to use in his inventions. The ancient cultures' practical and aesthetic application of different shapes and angles in the design of their homes, palaces, cathedrals, temples, and other buildings gave rise to sacred geometry, which became a way of life, philosophy, and religion for some people.

Builders of temples and churches who followed Pythagorean principles were able to erect massive edifices with colossal domes without obvious structural support, except for the very strict geometric relationship between each and every stone that allows them to defy gravity and not come crashing down. Not only are they structurally sound and architecturally eye pleasing, but they are also acoustic masterpieces, fully and distinctly carrying full voices and musical tones throughout their expansive prayer and meeting spaces. There is a certain ineffable feeling that comes over me when inside these especially constructed places of worship. It is a mixture of awe, humility, centeredness, and quiet joy. I very much doubt that I am alone in this regard.

The relevance of geometrical structures escaped notice by most scientists until 1980 when Robert Moon (1911–89), an engineer, physicist, and chemist at the University of Chicago, showed that the elements in the periodic table were based on these shapes.

Fig. 12. The two-dimensional representation of the icosahedron forms the "Flower of Life" pattern found in many ancient temples, mandalas, and artistic designs. Source: Wikimedia Commons.

Geometric crystallization is the way matter manifests in our reality, from the proton to the DNA, the sunflower, the nautilus, and a handsome face. Rock crystals such as diamonds (tetrahedral) and graphite (hexagonal) in the mineral world, the icosahedral self-assembly of proteins to form a virus, or the orthorhombic and rhombohedral crystal lattices of DNA amino acid proteins in the biological world are prime displays of geometric crystallization.

The ionized gas called plasma, which is the most energetic (i.e., highest-frequency) and only form of matter existing in the very hot conditions of the early universe and the core of stars, has a tetrahedral structure. Gas, the second-highest-energetic form, is octahedral. Liquids and then solids follow in descending order of frequency and are composites of the different platonic solids. Ice usually consists of clusters of fourteen molecules in tetrahedral arrangement. Blood hemoglobin is tetrahedral. Twenty clusters form an icosahedron, a structure that is stable in liquid water. Water clusters strengthen as they increase in size.

LOVE: UNITY BY NUMBER

Numbers unite together the two opposites of infinite sound and finite geometry. Infinite sound mathematically and harmonically results in the shape of matter. In the kabbalistic tradition and in its Hermetic and Pythagorean origins, "number" mediates between the infinite Creator and the finite material universe through harmonics. These traditions also associate the rhythmic beating of the heart with number and therefore, with the harmonic vibratory connection between the Creator and the created and vice versa.

Love is our linguistic expression of this one coherent vibration that binds all of nature, the music connecting everything to the Divine and to everything else, according to ancient written and oral teachings and traditions handed down through the ages by the Pythagoreans, kabbalists, Platonists, Muslims, and Christians. Love is fundamentally embodied in the esoteric teachings, beliefs, and practices of Eastern religions and mystics as well.

At the vibratory level, we are all one, united by the frequency of love. As Matthew 22:36–40 (KJV) relates: "Master, which is the great commandment in the law? Jesus said unto him, Thou shalt love the Lord thy God with all thy heart, and with all thy soul, and with all thy mind. This is the first and great commandment. And the second is like unto it, Thou shalt love thy neighbor as thyself. On these two commandments hang all the law and the prophets."

The Chinese philosopher Chuang-Tzu (also spelled Zhuang Zhou; 370 BC–287 BC) once said: "Behind the divisible there is always something indivisible. Behind the disputable there is always something indisputable. You ask what? The wise man carries it in his heart." We, people who possess wisdom beyond intelligence and acquired information, intuitively know that we have, in the very fiber of our beings, deep in the unfathomable bottom of our hearts, the pure vibration of love, the tune that allows us a certain knowing that is beyond comprehension and explanation, that connects us to the immutable source: the one and only original music, the singular frequency from which all frequencies emanate, the pure white light from which all colors of the rainbow shine.

ORDERING GROWTH AND FORM BY NUMBER

Leonardo Pisano Fibonacci (1170–1250) was born in Pisa, Italy, to a diplomat father, Guglielmo Bonacci. Fibonacci is short for *filius Bonacci*, Latin for "son of Bonacci." His father, who held a post in Bugia, North Africa (now Bejaia, Algeria), where he represented Pisan merchants who traded there, thought it would be a practical and useful move to have the young boy learn mathematics.

Fibonacci's studies were diverse. He was exposed to multiple cultural expressions of mathematics among the many merchants he met, since his father's assignments moved the family all around the Mediterranean region. It was in this environment, steeped in the accounting of money, goods, and services, that he learned to appreciate the convenience of the numerical symbols from 0 to 9 and the decimal place-value rule used by the Hindus and later adopted by the Arabs. He introduced the Hindu-Arabic calculation method to Europe and beyond through his 1202 book, *Liber Abaci*, published for the purpose of sharing with his countrymen in Pisa his newly acquired and useful method of accounting.

Fibonacci's dissemination of the decimal methodology was a precious gift to the world of accounting and mathematics that was appreciated then as it is now by most cultures for its ease of use, simplicity, and practicality.

The Fibonacci sequence of numbers, which can be found in the proportions of ancient Egyptian architecture (ca. 2600 BC–1 AD), was made famous by his brain-teasing mathematical question in *Liber Abaci*: "How many pairs of rabbits will be produced in a year, beginning with a single pair, if in every month each pair bears a new pair which becomes productive from the second month on?" The answer, which is the rabbits' reproductive sequence, goes like this: 0, 1, 1, 2, 3, 5, 8, 13, 21, 34, 55, and so on. Simply, it is derived by adding the previous two numbers to get the next one.

Fibonacci numbers appear in the sequenced arrangement seen in nature from tree branches to pine cones, "hen and chicks" succulent plants, and seeds such as those of the pineapple or strawberry. The tight packing pattern of the face of a sunflower is a showy arrangement of two flower spirals, going clockwise and counterclockwise. The total numbers of florets in each spiral appear in a sequence of 21 and 34 or 34 and 55 florets, or higher in bigger flowers.

The *phi*, or the golden mean or golden ratio (the ratio of length "a" to width "b" in a rectangle, or "a + b" to "a" equals 1.618034), is the ideal packing arrangement of petals and leaves for optimum exposure to the sun and the rain. The spiral growth and shape of seashells such as the nautilus (figure 13) maps out the golden mean as the growth of the mollusk logarithmically spirals outward in different degrees of rotation, for example 180 degrees or 360 degrees. The ratios of the length to the width at each and every step follow the sequential Fibonacci ratios of 1:1, 2:1, 3:2, 5:3, 8:5, 13:8, 21:13, and so on, all in close accord with the *phi*.

Fig. 13. A golden spiral unfolds in a cross-section of a beautiful nautilus pendant. Image from a personal collection.

In the January 2010 issue of the journal *Science*, researchers from Helmholtz-Zentrum Berlin für Materialien und Energie and the universities of Oxford and Bristol and the Rutherford Appleton Laboratory described quantum symmetry in cobalt niobate atoms lined up like magnetic bars only one atom wide. They found that when exposed to a magnetic field, the atoms could be tuned to approach Heisenberg's quantum uncertainty. In this state, the cobalt niobate atomic magnetic bar vibrates like a guitar string in the tiny scale of the quantum world.

They then discovered that the atoms magnetically resonated in a series of notes in perfect relationship with one another in exactly the golden ratio of 1.618. The same principle is at work when a soprano's voice achieves resonance with the atoms of a wineglass and the glass vibrates, reaching Heisenberg's uncertainty and weakening and dislodging resonant atomic connections within the crystal lattices, thereby shattering the wineglass.

Numbers and mathematics are real. Paul Dirac suspected the existence of the positron when his equations revealed a positively charged particle with characteristics exactly like the negatively charged electron. Computer manipulations performed according to mathematical laws give rise to an unceasing repetition of geometric or numerical patterns in any scale, called fractals. The same patterns repeatedly appear in fractals, similar in principle to the nesting concept of dolls and of tetrahedrons in spheres and spheres in tetrahedrons and so on.

The Mandelbrot set of fractal images, named after the Yale University professor of mathematics Benoit Mandelbrot (1924–2010), is created by feeding the result of a simple mathematical equation into the same equation over and over again, achieving increasingly complex and beautiful images that can go to infinity. Fractal images resemble the earth, snowflakes, lightning bolts, the branching of rivers and trees, and other familiar images.

Mathematics is fundamental to the appearance of all matter. Galileo said that Nature's great book is written in mathematical language. Einstein went further and asked a question: "How is it possible that mathematics, a product of human thought that is independent of experience, fits so excellently the objects of reality?"

While we can leave Galileo's statement at that and then at our leisure simply observe Mother Nature revealing her fidelity to mathematical truths, Einstein's question presses for a more deliberate attention and earnestly sober thought. The answer to Einstein's query lies in the question itself. His premise that mathematics is a product of human thought is faulty. Mathematics, rather than being a product of mental activity, has always been there, just awaiting discovery by resonant focused thought. The numbers themselves—their spatial relationships, their vibrational peculiarities, and their presentation in the faces of nature—have shaped humans' thoughts and have developed our abilities to discover mathematical patterns and their effects and ramifications. All we have to do is pay attention and discover them, just as poets, musicians, artists, philosophers, little children, and common people with unbridled imaginations have done.

All matter emerged from vibration through orchestration by number and geometry. Although form or geometry is bounded in perceived reality, the underlying frequency and harmonics are unbounded, eternal, and infinite. That means that the symphony materializing as your form or mine endures through eternity. Our unique personal eddy of electromagnetic frequencies—our music, our spirit—lives on.

8
Everything Vibrates

Nothing rests; everything moves; everything vibrates.
—The Kybalion

Phone calls are entertaining, if only because they are announced by all kinds of ringtones one cares to program in. My smartphone has Frank Sinatra crooning to me when it rings. Einstein imagined riding alongside a beam of light. Outdoing Einstein, Ol' Blue Eyes sang while sitting on a rainbow with the world on a string wrapped around his finger. What a life! He was in love.

It gets one thinking. The multiple colors of the rainbow from different vibrational frequencies of light, the watch that uses the resonance frequency of the quartz crystal (32,768 times per second), and love—they all have one thing in common: vibration. Everything vibrates. Everything has resonance. The ancients knew about it. From Egypt to China, harmonic music and the specific mathematical harmonic tuning of their musical instruments were strongly believed to connect humans to the divine and to the universe, affecting personal well-being in life and in the afterlife, as well as societal prosperity, harmony, and stability in times of peace, as well as success in acts of war.

Vibration was one of seven principles handed down orally to generations of followers of the ancient mystery school of philosophy of Hermes. According to *The Kybalion*, this school of occult wisdom was founded by Hermes Trismegistus, the purported author of *The Corpus Hermeticum*. He was believed to be a contemporary of Abraham's, the ancestor of the Hebrews and the Arabs who is supposed to have learned from Hermes. The authors of *The Kybalion*, identified only as the Three Initiates, wrote that the thrice-great Hermes was worshipped in Egypt as Thoth, the god of wisdom, and in Greece as the god Hermes. In both cultures, he was god of writing, wisdom, and magic and patron of alchemy and astrology. This master of masters prudently handed down his knowledge orally to only a select few who were unshakably devoted to the sacred philosophy and determinedly committed to utmost secrecy.

The term "hermetically sealed," which connotes a seal so tight that nothing, not even air, can get in or out, or a situation that is totally impervious to outside influence, is an allusion to a secret seal invented by Hermes that alchemists once used to make glass tubes airtight, and to the unambiguously principled secrecy of the Hermetic brotherhood.

According to the principle of vibration as taught by Hermes, the different vibratory rates of matter, energy, mind, and spirit account for their distinct manifestations. Spirit vibrates at an infinite rate of intensity and rapidity such that it is practically at rest, in the same way that the wheels of a rapidly moving vehicle look motionless. At the opposite end of the spectrum are forms of gross matter such as stones that have such low vibratory rates that they also seem to be completely at rest.

According to *The Kybalion*, "From THE ALL, which is Pure Spirit, down to the grossest form of Matter, all is in vibration— the higher the vibration, the higher the position in the scale." If we could understand this principle and utilize an appropriate method, we could control our own mental vibrations as well as those of others, including those of natural phenomena. According to Hermetic teaching per *The Kybalion*, "He who understands the Principle of Vibration has grasped the sceptre of Power."

The Hermetic philosophy is reported to have greatly influenced Pythagoras, whom we met earlier, widely considered the first philosopher. His disciples carefully studied and orally disseminated his teachings, which influenced Socrates, who taught Plato, who then shaped Aristotle's philosophy and, consequently, Western thought. Hermetic and Pythagorean philosophy, teachings, and practices are thought to have influenced the secret societies of freemasonry and rosicrucianism.

In search of truth, wisdom, and especially of occult knowledge from the gods, Pythagoras traveled far and wide and supposedly reached Egypt, Judea, Babylon, Arabia, India, Phoenicia, and probably many more places before he finally settled in Croton, an ancient city in southern Italy, a Greek colony then. He founded a select religious group, the Pythagorean Brotherhood, whose members strictly lived according to his secret teachings. He developed rites and practices for the community members, which they followed conscientiously. The brotherhood believed that the human soul is immortal, and to keep their souls pure, they led an ascetic way of life, practiced stringent personal discipline, and followed rigorous dietary restrictions.

Pythagoras experimented with many instruments and is said to have been an accomplished lyrist who developed a technique of tuning a musical instrument such as a lyre after he realized (according to legend) that the beautiful and harmonious musical tones from a local blacksmith's anvil were achieved by the simple mathematical ratios of the sizes of the hammers, which were two-thirds or one-half the size of another. Pythagoras went on to show that dividing the string of a musical instrument into sections with a 2:1 ratio produced a perfect octave interval, and a 3:2 ratio produced a perfect fifth (a frequency ratio of three cycles per second of the higher note to two of the lower note).

The three-limit "just intonation" (also called "pure intonation") system of tuning musical instruments based on these whole number ratios is the easiest to tune by ear and is the most consonant and, therefore, the most pleasing to hear. The musical instrument has to be tuned to the key of the music piece, and the music has to be played on that instrument in that key alone. In contrast, any music can be played on any key if the instrument has "equal temperament" tuning. The latter method, especially developed for the piano, has an octave of twelve notes with equal-frequency intervals, essentially forfeiting the frequency ratios of just intonation. As a result, the sound is no longer as harmonious to the human ear and is flat, lacking in color, and dissonant. (J. S. Bach's "Jesu, Joy of Man's Desiring" is best heard when the instrument is tuned using just intonation.) The just method and its offshoots—the meantone temperament and the well temperament types of tuning—were in use up until the twentieth century.

Pythagoras and his disciples believed that the whole universe was governed by simple mathematics and musical harmonics and that the movements of the stars, moons, and heavenly bodies were in accordance with mathematically spaced musical notes, played as if the whole cosmos were an enormous lyre making harmony: the "harmony of the spheres." He taught that "the heavens are the realm of pure number, where objects move in perfect, unchanging circles, the realm that can best be perceived through pure reason."

The Pythagorean theorem we learned in school (in a right-angled triangle, the sum of the areas of the two squares of the legs, a and b, equals the area of the square of the hypotenuse c) is attributed to him. It can be written as an equation relating the lengths of the sides a, b, and c as $a^2 + b^2 = c^2$. Because of the secrecy and oral nature of his teachings, no texts by Pythagoras seem to exist, and most accounts in history are by later writers.

THE HARMONICALLY VIBRATING UNIVERSE

Everything vibrates. A frequency can be recorded from everything in the universe, from the cosmic microwave background (CMB) to a member of the "rare earth" group of elements: neodymium

(Nd). Named after the Greek words *neos* and *didymos* or "new twin," neodymium has become a favorite in technology. Its amalgam with yttrium-aluminum-garnet crystal (Nd-YAG) emits a laser frequency that makes possible extremely fine and accurate surgical incisions. The material also allows highly precise diamond cutting—a novel, rare, and valuable technology in the diamond-cutting industry.

The alloy of neodymium, iron, and boron makes the strongest magnet ever, one that has found its way into the familiar world in the form of magnetic bracelets, watches, rings, oil filters, welding clamps, high-tech headphones and speakers, wind turbines, and the direct current motors used in hybrid and electric cars. When a mobile phone rings and vibrates, the vibrational frequency of this neodymium alloy magnet is often the source of the sound.

The noted Scottish botanist Robert Brown (1773–1858), peering through a microscope in 1827, observed the motion of pollen particles suspended in water to be rhythmic and musical, even though their individual movements appeared to be erratic. The motion (later known as Brownian motion) was what Einstein later described in his 1905 paper, although this time, the movements were of atoms suspended in gas. Einstein explained mathematically the mechanism by which the pollen particles in Brown's experiment were displaced here and there during their collisions with the water molecules and atoms in the suspension. The vibratory movement of the suspension as a whole was a function of harmonics.

In July 2014, a team of computer engineers and scientists from the Massachusetts Institute of Technology (MIT), joined by researchers from Microsoft and Adobe, published a report in the journal *ACM Transactions on Graphics* on the recovery of sounds from objects that have long been recognized to vibrate to sounds they are exposed to. Using a high-speed video camera, they recorded, without audio, the motions of an empty bag of potato chips and the leaves of a potted plant at the same time that an instrumental recording of "Mary Had a Little Lamb" played through a loudspeaker. By analyzing the vibrations of the bag and the leaves in the video and using high-tech computations, they were able to recover the original sound signals and then used them to reconstruct the audio of the instrumental rendition of the song.

Now, scientists have found evidence that the universe itself has its own vibration and has been oscillating, speeding up and slowing down seven times since the big bang 13.8 billion years ago, according to Harry Ringermacher and Lawrence Mead of the University of Southern Mississippi (figure 14).

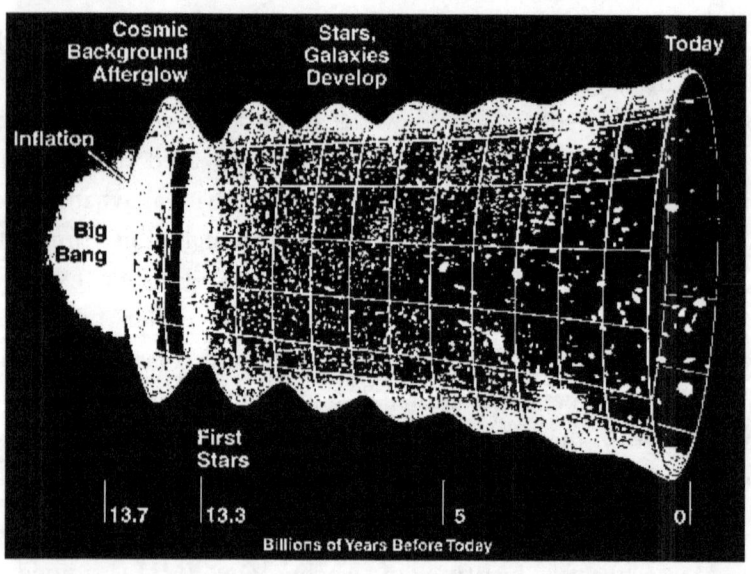

Fig. 14. A depiction of the seven oscillations of the universe over the last 13.8 billion years, dating from the big bang. Source: NASA / Wilkinson Microwave Anisotropy Probe (WMAP) Science Team

RESONANCE

Resonance, a term derived from the Latin word *resonare*, meaning to resound, has several definitions, depending on the context in which it is used. Pertinent to the topic of this book, resonance is defined as the tendency of a system to vibrate at its maximum amplitude at a certain specific frequency. Several extensions of the meaning proceed from this primary definition. The frequency at which a vibrating object achieves maximum amplitude is called its resonant frequency.

A legend about resonance tells of a young Galileo getting bored during a Catholic Mass celebration in a cathedral in Pisa in the late sixteenth century. Instead of paying attention to the service, he started watching the church chandelier as it swung back and forth. He noted that the chandelier maintained the same rhythm even when the distance it traveled varied from sway to sway. He synchronized its movement with his pulse and confirmed that it kept the same frequency, regardless of its swinging distance. This experience led to his interest in pendulums and their application to timekeeping devices.

Then he made a fortuitous observation that if he gave just a tiny push at the right time to a weight at the end of a string and then left it alone, the weight would keep on swinging back and forth for a long time. Galileo had discovered the principle of resonance.

Oddly, it was Christiaan Huygens—the prolific Dutch scientist discussed earlier, whose studies on the wave-like manner of light propagation preceded Young's double-slit light wave interference experiment—who noted that when two clocks were hung on the same wall, their pendulums would end up in the same rhythm. They would eventually become synchronized with one another. Pendulums are natural resonators, and he had a roomful of them to experiment on. He would start one swinging at a certain frequency, and even though all the pendulums swung at different intervals in the beginning, they eventually got entrained and synchronized somehow.

Resonant frequency is the reason why a guitar's A string spontaneously vibrates when someone plucks the A string of another guitar across the room. Another familiar encounter with resonant frequency happens with a playground swing. When force is applied on the swing in perfect timing with its natural rhythm, it swings higher, reaching its maximum height even if the force applied is uniformly maintained. The energy absorbed by the swing from the push is maximized when the force is in phase or "in resonance" with the swing's oscillation. A push at any other time would interfere with and dampen the resonance and its oscillation would be disrupted. The height the swing reaches would become lower or the swing would stop altogether.

A striking illustration of resonance is found in nature at the Bay of Fundy in Nova Scotia. A tidal exaggeration happens there with a whopping 16.7-meter (fifty-five foot) difference between low and high tides, as high as a five-story building. The reason is that the bay has a natural resonance, a rocking to-and-fro motion, just like the back-and-forth sloshing of water in a tub after it has been rocked several times. The bay is almost the ideal size, around 270 square kilometers (104 square miles)—not too short and not too long—just right for the frequency of the rocking motion from the mouth of the bay to its head and back to be thirteen hours.

The bay just happens to be in near-perfect synchrony with the oncoming rush of the tidal rise in ocean level created by the gravitational pushing cycle of the moon every twelve hours and twenty-five minutes. Just like the tiny nudge that maximized Galileo's weight-pendulum oscillations and the little push that makes the child's swing reach its maximum height, the timely tidal surge reinforces the natural sloshing of the bay's seawater from its mouth back to the shore and maximizes the water's force. To further amplify the tidal effect, the shape of the bay is such that it becomes narrower toward its head, and shallower too, from 130 meters to 95 meters (426 to 131 feet) deep, thus increasing the height the rushing water reaches at the shore.

To have a good grip on how remarkable the tidal resonance of bay of Fundy is, just consider the amount of water it moves in and out in a roughly 12.5-hour tidal cycle: 160 billion tons of seawater, more than the combined flow of all the world's freshwater rivers. The rapid rush of this much water creates mighty whirlpools and rapids. It is not a good idea to get caught swimming alone in the Bay of Fundy at high tide.

Expert manipulation of mechanical resonance is important in the construction of musical instruments, symphony halls, and cathedrals to optimize acoustics and to amplify rich sounds and clear voices. Medical imaging and treatment systems use magnetic resonance. Electronic devices such as radios, televisions, mobile devices, and GPS all utilize resonating parts. Living organisms, including the human being, use and depend on resonance to carry out their vital functions. This is something we must take heed of.

Nikola Tesla (1856–1943), a Serbian-American electrical and mechanical engineer, inventor, and physicist, was well known for his showbiz demonstrations of the effects of resonance, which brought him much fame as well as notoriety. He very seriously experimented on resonance and invented resonators, pocket-size devices that could make buildings shake. He actually created the effect of a small earthquake in Manhattan, sending panic-stricken neighbors to the streets and the police to his laboratory door. He nearly caused a skyscraper under construction to collapse just by holding his resonator against the steel frame of the building. His resonator also knocked a power plant offline in Colorado.

In a first-ever demonstration of robotics, Tesla floated a four-foot battery-powered unmanned steel boat with remote radio control. Tesla's wireless communication devices utilized the resonance of the sending instrument with the receiving device, each attuned only to the other, which rendered them impervious to the effects of other vibrations with non-resonant frequencies whirring about in the atmosphere.

VIBRATIONS IN BIOLOGICAL SYSTEMS

Vibrations. Resonance. They are present in all living organisms and are involved in every biological step integral to their existence and the propagation of their species. Vibration is vital to life in the plant and animal kingdoms. It is involved in all stages of plants' production and storage of energy, known as photosynthesis, the most important life-sustaining series of chemical reactions on earth, which harnesses light energy from the sun to convert water (H_2O) and carbon dioxide (CO_2) into energy-filled carbohydrate (CHO) molecules, the primary source of energy of the entire animal kingdom.

"Water is life's matter and matrix, mother and medium. There is no life without water," said the Hungarian-born biochemist and physiologist Albert Szent-Györgyi (1893–1986) of water. He received the 1937 Nobel Prize for his discovery of vitamin C (ascorbic acid). He also identified the chemical components of the Kreb's or citric acid cycle and their reactions in the body's metabolic function of energy production. Szent-György suspected that water exists in cells in an excited state and in a highly organized structure close to surfaces such as membranes and other water-loving cell structures such as proteins and nucleic acids. The highly ordered arrangement allows the water to start and promote sustained electronic excitation on everything it interacts with and to activate reactions and processes necessary for life.

Water organizes into crystal structures in its frozen states of ice and snow and, though not often thought of, it does so also in its ordinary liquid form at ambient room temperature under normal atmospheric pressure, most commonly in the simple but transitory tetrahedral form. Water droplets studied in five waterfalls in the Austrian Alps have been found to follow two ways of clustering. One group forms small aggregates of a few nanometers consisting of only a few hundred water molecules, and the other group forms large clusters containing millions of water molecules. The smaller clusters travel only a short distance, whereas the larger clusters are more stable and are able to travel much farther—up to several meters away from the waterfalls.

Stable crystalline supramolecular water clusters measuring tens of nanometers and even bigger—up to millimeters in diameter, consisting of millions to billions of water molecules—have been observed as well as photographed under transmission electron microscopy (TEM) and atomic force microscopy (AFM) in certain conditions. These water clusters are flexible structures and can be deformed easily by the tip of an atomic microscope probe.

Several researchers after Szent-Györgyi have worked hard on his posited quantum mechanism. Emilio Del Giudice, an Italian nuclear physicist, and his colleagues have shown that when a vacuum electromagnetic field interacts with water, the water molecules vibrate in resonance with the electromagnetic field. That results in the formation of large, stable areas of localized resonant vibrations between water molecules and electromagnetic waves, which are areas of standing waves called "coherence domains." These are observed in ordinary liquid water at ambient temperature and pressure. Water routinely fluctuates between its coherence-domain structure and its liquid form.

Water clusters help define the crystalline structure of other self-assembled supramolecular complexes and are affected by their surroundings. Water molecules are arranged in layers around the many structures they interface with—so much so that they become part of the structures themselves, allowing water to move energy as a continually evolving interconnected whole. When viewed under a polarizing microscope, the watery matrix appears like an ongoing fractal liquid crystal display (LCD), just like the plasma in an LCD television screen.

Although water is universally accepted to account for roughly 70 percent of the human body by weight, it is not commonly appreciated that the body, by number of molecules, is 99 percent water. As such, the human body is highly subject to the influence of electromagnetic radiation from its environment.

QUANTUM PHOTOSYNTHESIS

Photosynthesis is the key quantum performance of life. The addition of carbon (derived from carbon dioxide) to water results in energy storage in the form of carbohydrates and the release of oxygen into the air. The carbohydrates that members of the animal kingdom consume and the oxygen in the air they breathe fuel their bodies' energy production.

Photosynthesis is widely appreciated for its remarkably efficient energy utilization rate of over 95 percent, attributable to the quantum-mechanical and vibratory nature of light harvesting and processing. In contrast, only approximately 37 percent of the energy content of coal eventually ends up as electricity, the rest being lost as heat; similarly, the most efficient gas-fired generator offers only a 60 percent efficiency capability. The establishment of resonance between water and sunlight allows for effortless utmost economy of energy in the photosynthetic dance of life.

Photosynthesis occurs in two stages, both involving many complicated steps. The first stage, the "photo" part, involves the capturing of photons and the utilization of their energy. The "synthesis" part involves the production of carbohydrates (CHO) and their subsequent storage for the plant's use and for the members of the animal kingdom to harvest for sustenance of life.

How exactly a plant accomplishes this vital function has long been the subject of intense interest and research among botanists, biologists, and other scientists. Now it seems that an elucidation of the plant's secret mechanism is close at hand.

In 2014, a group of physicists from University College London, supervised by the Colombian theoretical physicist Alexandra Olaya-Castro, announced an irrefutable quantum experimental result showing that when a photon is absorbed by a plant, the photon energy creates an excitation of the electrons of the chlorophyll "a" and chlorophyll "b" molecules. Chlorophyll is the green pigment contained in the cellular powerhouses of the plant called chloroplasts, which are found in the green parts of the plant, mainly in leaves. Chlorophyll is one of a group of pigment-producing molecules called chromophores, which include the particular class of pigments called carotenoids that lend the red, yellow, and orange coloring to root crop, fruits, and leaves, especially autumn leaves.

Chlorophyll "a" maximally absorbs light with the specific wavelengths of 430 (blue) and 662 (red) nanometers, while chlorophyll "b" (blue-green) maximally absorbs light with wavelengths of 453 and 642 nanometers. Neither one absorbs light in the 500–600 nanometer wavelength range. After the transfer of energy from the photon to the chlorophyll "a" and "b" molecules, only the wavelengths in the 500–600 nanometer range remain unabsorbed and are reflected from the leaves to the eye's retina and to the brain, which interprets the wavelengths as the green color of leaves. The yellow, orange, and red carotenoids, such as those found in yellow squash, tomatoes, carrots, red pepper, and autumn leaves, are colored that way because they absorb light in the wavelengths of green to violet, reflecting back the yellow and red colors.

The hemoglobin of the red blood cells found in animals' blood carries oxygen from the lungs to the tissues. The hemoglobin has two parts: the "heme," which contains iron, and the "globin," which is a protein. The heme and chlorophyll share one thing in common: the "porphyrin" ring of atoms, a stable molecule that in heme has iron in the center and in chlorophyll has magnesium. Heme absorbs blue and green light and reflects red; therefore, one sees the intense red color of blood. The porphyrin ring serves an important role in the function of these two pigments. Electrons are perfectly able to move around this stable porphyrin ring freely, and so when they get energized, they simply jump out seamlessly from the ring into other molecules, transferring energy to them effortlessly. This process is how blood hemoglobin delivers oxygen to the tissues quickly and efficiently.

The biochemical reaction involved in plants' energy production and storage is actually very complicated, so we will only describe the simplest, most basic chemical steps here, which can be followed easily if taken slowly, step by step.

Upon the absorption of light, one photon excites one electron in one chlorophyll molecule, a process that boosts the energy level and the vibrational frequency of the chlorophyll molecular electron. This precise electronic excitation is transferred from one chlorophyll molecule to another by a quantum vibratory resonant mechanism—in a swift, wave-like motion—to reach the plant's "photosystem I" reaction center, where the "synthesis" part of photosynthesis takes place.

Within the reaction center, energy is transferred from the excited chlorophyll electron to an electron of photosystem I, which in turn gets excited and then gets kicked out of the reaction center. The expelled excited electron is accepted by the first of many specialized electron carriers and eventually reaches the photosystem II reaction center. The expulsion of the excited electron creates a deficit, an "electron hole" in the photosystem I reaction center. To replace the hole, photosystem II donates its own electrons to photosystem I, thus ending up with its own electron hole. Then, to satisfy its own electron requirement, photosystem II proceeds to snatch electrons from water (H_2O) by vibrationally exciting the electrons of the hydrogen component of water utilizing resonant sunlight, effectively freeing them from the hydrogen atom and thereby meeting its needs. This is Mother Nature's clever and elegant counterpart to the high-tech stripping of hydrogen electrons by Stodolna and others (discussed earlier) and to Einstein's thought experiment on the photoelectric effect.

With the hydrogen electron now stripped off, the rest of the water molecule is split into an oxygen molecule (O_2) and two free hydrogen ions ($H+$), which are the hydrogen protons without the electrons, separating them from each other. The oxygen molecule is then released into the air, making it available for harvest to support life in the animal kingdom.

The hydrogen ions combine with carbon dioxide (CO_2) from the air (by sharing electrons) to form carbohydrates (CH_2O), which naturally exist in several forms of molecular structures, each having a ratio of two hydrogen atoms to one each of carbon and oxygen. For example, the simple carbohydrate glucose molecule, written as $(CH_2O)_6$, has six carbon and six oxygen atoms to twelve atoms of hydrogen. Plant carbohydrates are the main sources of energy for the animal kingdom.

In a laboratory study published online in July 2014 in *Nature Chemistry*, University of Michigan researchers used short, ultrafast laser beam pulses in place of sunlight on spinach leaf protein complex extracts to intentionally start photosynthesis, and they took photos of the live transactions in real time. They were able to tell exactly when electrons were freed and successfully recorded spectroscopic signals that revealed vibrations during the electron-separation process.

What a spine-tingling adventure it is to peek into the fine-tuned secret operations of Mother Nature that give us life.

QUANTUM VIBRATIONS AND THE BUSINESS OF LIVING

Plants utilize vibrations in other functions besides photosynthesis. Acoustic waves in the ultrasound range and electromagnetic radiation in membranes activate the upward flow of water in plants and remove gas bubbles (which form in the sap) to prevent obstruction of the flow of water.

In a 2009 study reported in the journal *Applied Physics Letters*, a group of Spanish scientists cleverly and non-invasively recorded, live, the ultrasonic voices of plants from their leaves. The team used portable ultrasound (US) machines to emit sound frequencies that then reached the plants through air vibrations. The leaves echoed back resonant frequencies, which the US machine picked up and recorded, revealing the plant's state of tissue hydration, density, and thickness—without so much as a touch.

Plants possess a mechanism of vibratory sensory reception, the nature of which is still being defined, which they effectively use to detect gravity, light, temperature, and touch—environmental cues that undoubtedly affect their behavior. One may reasonably suspect this ability to involve "microtubules," which plants possess in common with all organisms with a cellular membrane (discussed shortly in the section of this chapter titled "Vibrating Microtubules").

The mysterious Venus flytrap, aptly named after the Roman goddess of love, snaps its trap shut to securely embrace its prey in a matter of 0.3 seconds, faster than one can blink an eye. The Venus flytrap has a fascinating and strangely sophisticated technique for survival. Despite being able to derive its nutrients from the soil, water, and air, just like any other plant, its nutrient-poor natural habitat in hot, humid, and boggy marshes forces this plant to obtain its food from the animal kingdom, a reversal of the natural order of things.

For this very purpose, nature has provided the plant with a trap consisting of a specialized upper portion of the leaf called lobes, located on each side of a central leaf midrib, to which the lobes are attached in a hinged manner to allow them to open and close. The top surfaces of the lobes are convex in shape when the trap is fully open and are equipped in the center with three rigid trigger spines, or hairs, that are acutely sensitive to pressure. The lobes are colored red with the pigment anthocyanin, and they secrete nectar for the specific purpose of attracting insects. Their outer edges are lined with a row of long, curved spikes that interlock upon closure of the trap to prevent the insect from escaping.

The cells of the lobes are divided into distinct upper and lower layers, each connected to the other by specialized pores. Their function is of a hydraulic nature, flawlessly coordinated for quick and efficient response in the act of trapping its prey. When the trap is open, the top layer's hydraulic pressure is higher than the bottom layer, making the top cells more turgid (i.e., swollen) than the bottom cells, thus allowing the top layer to curve over the softer bottom layer, which explains the lobes' convex shape.

When something such as an insect (or any other object) brushes against two or more trigger spines, or when a single trigger spine is bent more than once, the mechanical pressure on the spine is converted into an electrical signal. The electricity spreads in a wave-like manner toward the midrib and, in so doing, incites the specialized pores connecting the two hydraulic cell layers to open. Water rushes from the leaf's top layer cells to the bottom, quickly dropping the hydraulic pressure of the top cells. This precipitous shift in pressure creates an imbalance in the turgidity and elasticity ratio between the top and bottom layers. The moment a critical geometric ratio between tissue turgidity and elasticity is reached, the lobes instantly snap back, inverting the two lobes from convex to concave and abruptly closing the trap, which then functions as a pouch with Ms. Venus's desired protein fare inside.

At first, the trap is not quite tightly shut; a tiny opening remains to allow smaller creatures to escape. After some time, aided by the captive's wiggling, the trap closes tightly on its live main entrée, and the dining begins. The secretion of digestive enzymes into the stomach-like pouch enables the plant to process and absorb its prey. If what the Venus flytrap has caught is not food, then the plant just spits it out a little after twelve hours, when the lobes unfurl again into their turgid convex shape, its trap prepared once more for another shot at luring a proper meal. Otherwise, a live insect takes between five to twelve days to be fully digested and absorbed and for this marvelous trap to be groomed and ready again for Ms. Venus to undertake her next intriguing carnivorous seduction adventure.

Plants respond to music. Indian polymath, physicist, biologist, and botanist Jagadish Chandra Bose (1858–1937), who studied physics at the University of Cambridge, is now widely recognized as the first to invent the wireless detection device and the millimeter-band radio. He also had the rare honor of having a crater on the moon named after him. In his studies on plants, Bose discovered that they grew faster when exposed to soft and enjoyable music. He also observed differences in the plant membranes' electrical activity when exposed to soft speech or pleasant music versus stressful harsh words or loud music.

Plants have long been known to communicate with one another to help them recognize family members (in order to avoid untoward competition among kin) and to promote cooperation and problem-solving as a group for the survival of their species. Botanists have shown that plants sacrifice themselves, in true altruistic fashion, for the sake of their relatives or community. The mechanism by which they exchange information was once thought to occur only through the environmental cues of gravity, light, temperature, or touch, or via biochemical signaling through the release of volatile organic compounds into the air and through the root network of underground vegetative fungal connections. It is already accepted in the scientific community that plants talk to one another through quantum vibratory communication.

In a study by the evolutionary ecologist Monica Gagliano and her colleagues, the researchers noted that the roots of corn kernels suspended in water grew toward the source of sound at a frequency of 220 Hz transmitted through the water. This sound was in the same range as the frequency of the structured spikes of spontaneous vibrations that these roots were recorded to have emitted. The corn roots were attracted to the source of the sound of resonant frequency.

In another study, Gagliano and a team of colleagues studied the behavior of germinating chili seeds and chili seedlings. When the seeds were placed beside basil plants (known by gardeners to be beneficial companion plants), their germination rate was high: even higher than when grown with adult chili plants. When planted with fennel (known to be a non-beneficial bad neighbor), the volatile chemicals from the fennel hindered the germination of the chili seeds.

When the researchers blocked all forms of communication—whether by contact, light, or chemical signaling above and below ground—they found that the seeds and seedlings could still identify relatives and discern neighbors separately. The young plants accordingly modified their growth patterns, as though their channels of communication had never been blocked at all. Specifically, with all known means of communication blocked off, the chili seeds still detected the presence of the fennel plant and, surprisingly, germinated much faster.

According to Gagliano, "Because germination rates have lifelong fitness consequences, selection should clearly favour mechanisms allowing a plant to detect its neighbours and hence its forthcoming competitive environment and regulate its developmental responses accordingly at the very onset of its life (i.e. seed stage)."

VIBRATIONS IN THE ANIMAL KINGDOM

All members of the animal kingdom utilize vibrations extensively. Their world appears to be suffused with a wide range of low frequency vibrations that we humans are barely aware of. Animals elicit vibrations by the use of their vocal chords and their whole bodies (tremulation), by rubbing their body parts (stridulation), and by drumming or stomping on the ground. They have a somato-sensory or whole-body mechanism for receiving vibratory signals that are then mechanically transmitted to the brain directly, and indirectly through the ears for regular hearing reception.

Animals use vibrational signaling to communicate with their family members as well as members of their communities and of other species. They use vibrations to mark their territories, to attract mates, to find prey, or even to cunningly attract potential prey by imitating the prey's very own frequencies.

Earthworms have a predictable response to the vibrations of moles digging holes in the ground to catch them for food: they wriggle their way out of the ground when they perceive such vibrations. To the misfortune of the worms, other predators, such as birds and turtles, have in turn learned to emit vibrations that imitate the digging sound of moles; they send those frequencies through the ground, which then forces the poor worms out.

To make things worse for the hapless earthworms, humans who use them as fishing bait or who simply catch them as a sports activity likewise have a technique called "worm charming" to harvest them without so much as touching the ground with a shovel. Worm-charming methods of vibrating the soil include "worm grunting," in which a wooden stake called a stob is driven into the ground and rubbed with a "rooping iron" to make a grunting sound, and "worm fiddling," in which a dulled saw is moved across the top of the stob. In any case, the worms don't stand a ghost of a chance.

In a 1925 study, a Slovene biologist named Ivan Regen studied insect sounds by positioning a male cricket behind a microphone and female crickets in front of a loudspeaker. When the male cricket made sounds, the females moved toward the loudspeaker and not to the male behind the microphone, indicating that the signal was mediated through sound vibrations.

The sense of smell is mediated through molecular vibrations. Fruit flies distinguish between two identically shaped molecules that emit different vibrations and different scents, thus nullifying the theory of an anatomical lock-and-key fit between the receptor and the stimulant molecules. Dolphins communicate with sounds created by vibrations of the nasal cavity tissues, not of their respiratory air column, which gives them the advantage of sending vibrations through water from any depth—a more efficient way than if the vibrations were transmitted through air.

VIBRATING MICROTUBULES

Inside any living cell with a sheath or membrane, including that of the single-celled paramecium and the complex organism called a human being, is found a byzantine network of tiny microtubules, measuring about 25 nanometers in diameter, originating from a microtubule-organizing center near the cell nucleus. These nanotubes are assembled from specific proteins called tubulin polypeptides (chains of amino acids bonded together in a special sequence to make proteins) that self-arrange into long, thin strands called proto-filaments. Thirteen proto-filaments come together, with eleven of them forming a wall encircling the last two. Each microtubule is filled with highly structured molecular water clusters specially arranged in a unique manner after the Fibonacci sequence of 3, 5, 8, and 13.

The half-life of a microtubule is around ten minutes or so. Microtubules undergo constant self-directed structural assembly and disassembly on both ends, growing slower at the originating end, which is called its "minus end," and faster at the extending end, called the "plus end." The plus end connects with the inner surface of the cell membrane and other parts of the cell, including the nucleus and its DNA content. Specially and critically assembled to form the cytoskeleton (cell skeleton), microtubules crisscross the cell's cytoplasm like an intricate network of superhighways for the purpose of delivering information and transporting water, vesicles, granules, nutrients, chromosomes, and mitochondria. The cytoskeleton determines the architecture of the cell and stabilizes it.

The microtubular plus ends attached to the cell membrane grow in such a way that a specially arranged circular collection of them form tubular protrusions of the membrane that look like hairs around 1–10 microns (micrometers) long called cilia, and flagella, a few microns longer. An extreme exception to this size is the length of microtubules in a nerve cell axon (the part that looks like a very long thread of a tail), which can reach one meter in length. Under a light microscope, cilia and flagella look like hairy projections and exhibit rhythmic sweeping motions in unison. This wave-like motion is evident in the lining of the lung bronchial passages (see figure 15), which move mucus secretions and foreign bodies upward, to be cleared by the reflex coughing mechanism.

Linked with the microtubule tubulin polypeptide complex are proteins called microtubule-associated polypeptides (MAPs) that behave in concert with the microtubules. At the microtubule plus end, a MAP with a particularly high molecular weight has been found to resonantly integrate with special DNA polypeptides, called kinetochores, during cell division. This unification is highly functional in nature and is crucial to the delicate disentangling and accurate separation of chromosomes.

When a critical DNA-to-MAP ratio is reached, the DNA then behaves synchronously as a part of the microtubule water-tubulin-

MAP-kinetochore complex, the whole thing acting as one molecule for the maximally efficient reception and transmission of vibrational information, now recognized as a necessary feature for precise cell division and most essentially, conscious perception.

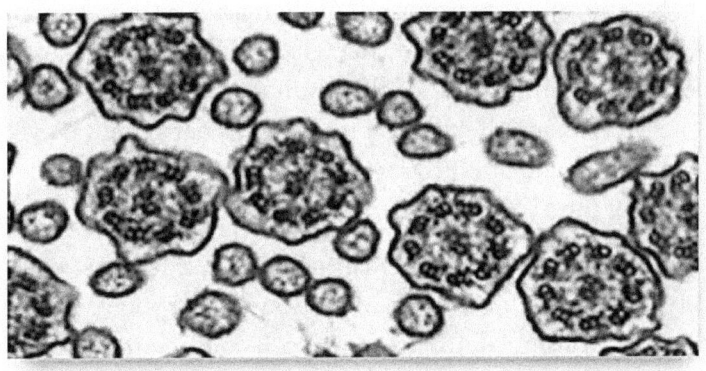

Fig. 15. Electron microscopic image of cilia cross-sections from mouse trachea, each showing nine doublet microtubules close to the membrane and two doublet microtubules in the center. Source: National Heart, Lung, and Blood Institute (NHLBI) / National Institutes of Health (NIH).

The vibrations of specially arranged atomic water molecule clusters inside the microtubular nanotubes initiate quantum resonance in the kHz and MHz range frequencies in the surrounding tubulin proteins, such that the microtubule and its surrounding protein structure behave as a single molecule, regardless of length or size. In this sense, the resonance of the molecular water controls the optical and electronic behavior of the microtubule, and thus its function, which justifies Szent-Györgyi's hypothesis that the highly structured excited water layers close to the surfaces of cellular membranes, proteins, and nucleic acids play a key role in the electromagnetic reactions that underlie the biological processes of life.

VIBRATION AND HUMAN HEALTH

Circa diem. Around a day. Circadian: a rhythmic cycle that rules the biology of all light-sensitive organisms, including humans, that possess clock oscillators inside the cells of their bodily organs. These oscillators are independent of any external cues except for sunlight and are synchronized by a master clock, a pacemaker called the suprachiasmatic nucleus (SCN) located inside the hypothalamus at the base of the brain above the optic chiasm (the eye-nerve crossing) behind the forehead in between the eyebrows. The pineal gland, often thought to be the "third eye," sits toward the back of the SCN.

The pacemaker, through quantum light oscillations, controls all bodily processes, including digestion, absorption, elimination, body metabolism, detoxification, cell replacement and repair, immune function, blood pressure, heart rate and rhythm, kidney and adrenal gland function, and the sleep-wake cycle. Disorders such as jet lag and metabolic syndrome X—marked by diabetes, high blood pressure, elevated insulin and triglycerides, obesity around the waist, and low HDL (or "good,") cholesterol—involve disruption of the circadian pacemaker rhythm.

Just as sunlight starts an electronic excitation in the process of photosynthesis, it likewise excites the structured crystalline water inside the microtubules of all living creatures, starting the biological molecular interactions necessary for life to roll vibrantly along. This could be the mechanism by which exposure to sunlight makes a person feel generally healthier and happier, vitamin D levels increase, and seasonal affective disorder (SAD) and low moods or depression either improve or disappear altogether.

For example, healing waters from the Ganges River and from Lourdes (a grotto and pilgrimage site in France) have been found to have the characteristics of the structured water found in cells. Under the auspices of the National Institutes of Health, scientists reviewed various Lourdes miracle cures and came to this conclusion: "The least that can be stated is that exposures to Lourdes and its representations (Lourdes water, mental images, replicas of the grotto, etc.), in a context of prayer, have induced exceptional, usually instantaneous, symptomatic, and at best physical cures of widely different diseases." Resonance between the structured molecular water clusters in the environment and in human body cells may be one of the reasons why simply being around water, especially waterfalls, imbues one with a feeling of wellness and vitality.

Acupuncture is widely accepted in the Western practice of medicine, mainly for pain relief in the United States, and has been the subject of extensive study. The practice is based on the presence of a network of cells sensitive to electricity and light distributed along certain pathways or meridians, which correspond to different organs of the body. Energy, or *qi* (also spelled *chi*), from the universe runs through this meridian network, which is similar to the concept of *prana* flowing through the *chakras* in Ayurvedic medicine.

Blockage of this electromagnetic energy network causes discomfort or ill health. According to the practitioners of Eastern medicine, stimulating the acupuncture points along the meridians with extra-fine needles or electrical probes allows the energy from the universe to flow through these energy channels, thus rebalancing the body's energy field and bringing it back to health.

In the 1940s, German physician and engineer Reinhard Voll found that different areas of the body surface showed much higher conductance than the rest of the body, which he found to correspond to the Chinese acupuncture meridians or channels. He invented the "electro-acupuncture by Voll" (EAV) machine to measure the electrical flow through this meridian network. Today, the EAV has been computerized and is popularly used (more so in Europe than in the United States) to identify blockages of energy flow that correspond to different organs of the body.

The earth emits a non-stop droning of its own. The "hum" has brought people headaches, anger, low moods, and many a sleepless night—phenomena variously ascribed to tinnitus, leaky pipes, wind farms, and many other sources of noise. In a 2015 study published in the journal *Geophysical Research Letters*, seismologists found a plausible explanation for this situation: the whole earth rings. The researchers found that microseismic vibrations of 3 to 300 seconds at a time are created by ocean waves as they roll over the ocean floor and collide with ocean ridges, underwater mountains, and continental shelves and may explain most of the hum.

Researchers' recognition of the health effects from radio frequencies of 3 kH to 300 GHz is based on a body temperature rise of more than one degree Celsius; the biological effects of elevations less than that have not been identified or adequately studied as of this writing. Extremely low frequency (ELF) exposures below 300 Hz (household electricity is 60 Hz) have been implicated in chronic fatigue syndrome and disruption of the immune function. The World Health Organization's International Agency for Research on Cancer has classified these radiofrequency fields as "possibly carcinogenic to humans."

While large-scale population studies on the effects of cell phones on cancer rates have not been conclusive, the utilization of sugar has been found to increase on the side of the brain most exposed to cell phone antennas, which reflects an increase in metabolism of tissues in that area—an accepted predictive marker of cancer activity.

Furthermore, Anirban Bandyopadhyay, senior scientist at the National Institute for Materials Science in Tsukuba, Japan, and his colleagues, have revealed that fractal frequencies of the brain microtubules are in the 100 kilohertz to terahertz range and certainly would be affected by the electromagnetic frequencies of mobile devices with emissions in the 100 kHz to 300 GHz range.

Wi-Fi channels, Bluetooth devices, and smartphones are in the microwave range of 1 mm wavelength (300 GHz frequency) to 1 m wavelength (300 MHz frequency). Only time and further research will tell in what way and to what extent living cells are affected by these artificially induced electromagnetic frequencies.

The discovery of the quantum wireless mode of communication between brain microtubules has only recently overturned the long-held neurochemical approach to brain functioning, and the method of investigation in which frequency fractals are used is still new. In contrast, the body has long been known to be mostly water; as noted earlier, by number, the body is 99 percent water molecules. Water, in its existence as clustered molecular structures inside the cells of organisms, vibrates in resonance with an oscillating electromagnetic field, whether inside or outside the human body.

The strong magnetic fields used during MRI scans for diagnostic purposes fall within the ranges that the human body is exposed to in its usual environment, and so most medical authorities deem the harm to be much less than the risk posed by ionizing X-ray imaging, including CT (computed tomography) procedures. According to the National Cancer Institute, one typical CT scan represents a 1 in 2,000 extra risk of fatal cancer. Physicians (including radiologists) and patients should, therefore, carefully and thoughtfully choose whether to opt for a CT examination. In 2011, the European Union banned the usage of backscatter X-ray security scanners for reasons of medical safety. The US Congress banned their use in 2012 owing to public uproar because, oddly enough, of the embarrassingly graphic images the technology produces and over concerns for the privacy of information storage.

The United States Transportation Security Administration (TSA) removed such scanners from all airports in 2013. More recently, the TSA has deployed millimeter-wave scanners, which are microwave or radiofrequency non-ionizing radiation scanners. It is important to note that this scanner uses a frequency range of 24 to 30 GHz, only minimally higher than the 19.98 GHz vibration of the DNA helix backbone; the millimeter-microwave scanner and DNA could thus be said to be nearly resonant.

The vibration of the ubiquitous household microwave oven (2,450 MHz) is in the microwave range. During ordinary use, microwave ovens have built-in protection from leakage, but the TSA's millimeter scanning is, in effect, an intentional exposure to microwave radiation.

The long-term effects of repeated or prolonged exposure to millimeter-wave radiation on human health from TSA scanners were not fully determined before they were deployed in airports and other high-security places. All we have is a reasonable suspicion of an adverse effect of millimeter-wave exposure on resonant living cell microtubules based on a dramatic analogy to the shattering of a wineglass by vibrational dislodgement of its crystals in response to a soprano's resonant voice frequency.

HEALING BY RESONANCE

As a physician, Christian Friedrich Samuel Hahnemann (1755–1843) became dissatisfied with the way medicine was practiced in Germany, which included bloodletting. He gave up his practice in favor of medical writing, translation, and research. In translating the Scottish physician William Cullen's book *A Treatise on the Materia Medica*, he came across the treatment of malaria with cinchona bark, the effectiveness of which was attributed to its astringency, now recognized as due to its quinine content. Finding that other astringent substances were not effective in treating the disease, Hahnemann ingested cinchona extract and experienced symptoms of malaria. He postulated that a substance that brought on symptoms of a disease in a healthy individual could alleviate the same symptoms in someone with the actual disease. He further found that diluting the substance and shaking it vigorously reduced the side effects of the potion and increased its effectiveness. As he concluded, "that which can produce a set of symptoms in a healthy individual, can treat a sick individual who is manifesting a similar set of symptoms." Thus was homeopathy founded as a medical practice.

Practitioners of homeopathy (homeopaths) use substances that have been increasingly diluted in alcohol or water and repeatedly and vigorously shaken (a technique known as succussion), such that the final product no longer contains even a trace of the original material. Homeopaths maintain that a transfer of energetic information occurs with serial dilution and succussion.

That the mode of action of homeopathic preparations is based on resonance frequency was confirmed in a study by the German alternative medicine practitioner Karin Lenger and her colleagues. Using two Tesla coils (briefly described in the next chapter) to generate high-frequency longitudinal waves, the researchers showed that homeopathic high potencies with frequencies in resonance with the coil frequencies attenuated the magnetic field present around the coil. They identified the different homeopathic potencies as magnetic photons, and they concluded that "each homeopathic potency must have its specific frequency spectrum and its specific energy, which is the homeopathic information."

COHERENT RESONANCE IN OSCILLATING BODY SYSTEMS

In a healthy human, a pattern of oscillations in the heart rate occurs with breathing at ten-second intervals (0.1 Hz), which is the cardiovascular resonant frequency. Arterial blood pressure oscillations called Mayer waves (named for the work of German scientist Siegmund Mayer and colleagues) also occur at ten-second intervals. Variations in the beat-to-beat interval between heart beats (taking time and breathing into consideration), called heart-rate variability, are thought to be reflections of heart-brain interactions and the dynamics of the autonomic nervous system: the sympathetic (flight-or-fight) and the parasympathetic (rest-and-digest) systems.

Cardiac coherence is said to exist when the heart rate becomes more ordered and there is a relatively harmonic sine wave–like signal of 0.1 Hz or 10 second duration. Researchers have found that the greater the variability of the heart rate, the longer the life expectancy after temporary or prolonged stressful events, such as emotional trauma, heart attack, heart failure, the condition known as broken heart syndrome, posttraumatic stress syndrome, depression, and hypertension. These conditions, all of which are characterized by low heart-rate variability, are associated with inflammation, cortisol elevation, and heightened sympathetic and lowered parasympathetic nervous system activity.

Praying the Ave Maria in the original Latin priest-parishioner responsorial cadence achieves a perfect ten-second breathing rhythm, resulting in the enhancement of Mayer waves. English, Italian, and French translations of the Ave Maria do not result in the same rhythm and have no effect on the Mayer waves.

Fundamental vibrations known as Schumann resonances are present and have been recorded in the layers of atmosphere from 85 to 595 kilometers (53 to 370 miles) above the surface of the earth (see figure 16), They were discovered by German physicist Winfried Otto Schumann (1888–1974) in 1953. These vibrations are normally not discernible, and it takes an excitation by a lightning strike for them to be recorded. The strongest resonant frequency is 7.83 Hz, close to that of a standing human, which is 12.3 +/–0.1 Hz.

The resonant frequency of the earth atmosphere is important in regulating the body's circadian biological clocks controlled by the master pacemaker, the SCN, located in the base of the brain. The human body tunes in to the music of the earth and gets entrained with the earth's intrinsic rhythm.

Fig. 16. The atmospheric layers of the earth as viewed across earth's landscape from the International Space Station. Source: NASA.

Even though the whole earth has zero electrical charge, its surface charge at sea level is negative, thanks to the negative electrons deposited by lightning and thunderstorms. When a person is in bodily contact with the earth, electrons from the earth flow into the body, which is a good conductor. The electrical potential of a person's body changes to approximate or equal that of the earth.

In a lecture on electromagnetism, Nobel laureate Richard Feynman (1918–1988), a brilliant American theoretical physicist, said that when the electric potential of the human body approaches or is the same as the earth's, it is in unity with the earth's potential and the earth acts to cancel the electromagnetic fields that are foreign and harmful to the body. The human cells are, in effect, "grounded."

Much in the same way that electrical grounding protects houses and appliances from electrical surges and lightning, the popular wellness practice of "earthing" or "grounding," which optimally requires direct earth-to-skin contact while walking barefoot or lying on the ground, tunes the electrical activity and frequency of the body to that of the earth. Earthing calms down the body's unnecessary and harmful electrical activities that cause ill health and has been found to improve heart rate variability, help chronic fatigue symptoms, ease various aches and pains, and promote a sense of wellbeing, vigor, and vitality.

The rebalancing effect of grounding on the body's electrical charge may well explain why it feels so heavenly to be lying deep in the warmth of a sandy beach on a balmy afternoon when the sun is easing ever so softly into the hush of dusk.

9

Electromagnetism

All beings of the universe are so connected that it is as if they touched each other.

—*Franz Anton Mesmer*

Benjamin Franklin (1706–90) cuts a giant figure in American history as one of the founding fathers of the United States of America and as a member of a committee of five selected by the Continental Congress to draft the Declaration of Independence in 1776. He was a talented writer, scientist, and inventor. He became famous for publishing the widely circulated *Pennsylvania Gazette* and the popular *Poor Richard's Almanack* from 1732 to 1758 (Richard Saunders was his pen name).

Many of Franklin's clever inventions were practical objects that still find use today, including the Franklin stove, the flexible urinary catheter, bifocal lenses, the lightning rod, and many others more. Despite all his achievements, he is dramatically enshrined in people's imagination as a grown man flying a kite in a storm and getting an electric spark caught by the kite from the cloud and conducted through the twine he was holding.

As the story is told, Franklin performed the experiment by constructing a silk kite with two crossed sticks, one longer than the other. Attached to the top part of the longer stick serving as the head of the kite was a foot-long (30 cm) wire to attract electrical current. Tied to the junction of the two sticks was a long wet hemp twine, on the free end of which was attached a metal key and a silk string to hold the kite down with. With the assistance of his young son, Franklin went out to an open field in a thunderstorm to fly his experimental kite, taking cover in a shack. It took some time before the stormy sky cooperatively sent an electrical charge down to the metal wire of his kite, then to the twine, rendered more conductive by its wetness. The fibers of the twine stood on end as the electrical charge traveled down to the metal key, from which Franklin got a spark of electricity once he put his finger near it. The electrical charge did not travel all the way down the silk string, the free end of which was held and kept dry by Franklin inside the shack.

Franklin's experiment attracted international attention, both in the scientific community and among the general populace. His newly gained fame, political stature, and social charm are credited with having been instrumental in securing the support of France for colonial America's struggle for independence.

FROM RAGS TO OTHERWORLDLY RICHES

With scanty formal education to build on, Michael Faraday (1791–1867) rose to great scientific heights through his own research and native genius, and he gained a reputation as one of the most influential scientists of all time. The son of a poor blacksmith in South London, he received minimum formal schooling. Out of necessity, he quit school and started working at the tender age of fourteen as an apprentice to George Riebau, a bookbinder and bookseller, who became his first mentor, and whose books provided him with further education. Riebau was impressed by Faraday's notes and commentaries on lectures delivered by John Tatum, the founder of the City Philosophical Society. He alerted William Dance, an English music teacher and cofounder of the Philharmonic Society of London, to the boy's talent. Dance, touched by the extraordinary intellect in someone so young and so poor, and aware of the lad's growing interest in science, saw to it that Faraday got tickets to attend lectures by the English chemist Sir Humphry Davy, a member of the Royal Society and the director of the Royal Institution's chemistry laboratory.

It was during these lectures that Faraday's appetite for scientific explorations got finely whetted. The voluminous notes he took, which he bound into a book, became instrumental in securing him a position as secretary to Davy when a laboratory accident damaged the director's vision. A chemical assistantship position opened at the Royal Institution in 1813, and Davy offered the job to Faraday. In this capacity, he contributed much to the work of Davy and other Royal Institution scientists; for example, he and Davy co-invented the miner's safety lamp in 1815.

Faraday experimented on electromagnetism, and it was in 1831, while he was studying the principle behind the workings of an electric motor, called electromagnetic rotation, that he came upon the phenomenon of electromagnetic induction. This is now known as Faraday's law of induction, which states that "a changing magnetic field creates an electric field."

The principles of electromagnetic induction and rotation eventually led to Faraday's development of the first electric motor, which converted electric current into useable mechanical energy, a conversion on which many modern conveniences are based. He further facilitated the advancement of the Industrial Revolution by the numerous types of electrical hardware he invented, as typified by the electric generator, the transformer, and the Faraday cage, the last of which is still used to protect electronic equipment from lightning strikes and shield them from electromagnetic interference.

The term "farad," was given in his honor to a standard unit of measure which indicates the capacity of a system to store an electrical charge under the International System of Units (known as SI units from the French *système international*). His interest in chemistry led to his invention of a prototype of the Bunsen burner, which is still used in chemistry labs today, and his usage of the terms "ion," "electrode," "anode," and "cathode" eased their incorporation into the vocabulary of modern-day chemistry.

Since he believed that nature was an interconnected whole, having come forth from one source—a God that was one with His creation and whose creation was one with Him—he believed that electricity and magnetism, being forces that belong to an interconnected nature, were also interlinked. He then sought the experimental connection between electricity and magnetism, and even gravity. Faraday thought that every scientific theory, no matter how elegant or sophisticated, was ultimately subject to judgment by objective reality.

Overcoming the obstacles thrown across his scientific track by his lack of formal education and his lower-class social background, Faraday became the very first Fullerian Professor of Chemistry, a lifetime position at the Royal Institution in London. He declined the offer to become its president, however, and we would be calling him Sir Michael Faraday today had he accepted the royal offer of knighthood. As a conscientious follower of the austere religious way of life of the Sandemanian Presbyterians, he could not welcome the fame and glory that would come with the presidential position and knighthood. Humbly, he accepted a modest pension and a comfortable retirement home from the British government.

Michael Faraday's strong relationship with God gave him the courage and strength to refuse a British government request to make poison gases for the Crimean War of 1853–56. Professor Michael Faraday, the scientist and experimentalist, and church elder Michael Faraday, the upright and devout Christian philosopher, were strictly compartmentalized. Even so, the dedicated and stringent approach he applied in one discipline was paralleled in the other, and his spiritual convictions illumined his scientific pursuits.

MAXWELL'S ELECTROMAGNETIC CONNECTION

"The beauty of electricity and of any other force is not that the power is mysterious and unexpected, but that it is under law and that the taught intellect can even now govern it." Those were Faraday's words when he was searching for the experimental connection between electricity and magnetism in order to forge their practical application. James Clerk Maxwell, who learned about electric field lines at age twenty-four when Faraday's scientific paper was read in 1855 to the Cambridge Philosophical Society, was exactly that schooled intellect needed to work on electromagnetic induction.

Maxwell worked on Faraday's model and developed a set of equations that clarified and linked the knowledge about electricity and magnetism. He used four different equations expressing laws for electricity and magnetism developed by three scientists: Carl Friedrich Gauss, Faraday, and André-Marie Ampère. Maxwell's unified equations revealed that a changing electric field creates a changing magnetic field, and, in turn, a changing magnetic field creates a changing electric field.

From the four equations and his calculations, Maxwell found the speed of propagation of the electromagnetic field to be equal to the speed of light: 299,796 kilometers (186,284 miles) per second (kps or mps) as measured in 1862 by Léon Foucault: only four kps more than the twentieth-century laser-acquired figure of 299,792 km/s in a vacuum. Maxwell established that light travels as waves and that all propagating waves travel at the speed of light. Light, he concluded, is a propagating electromagnetic wave. His electromagnetic theory was further clarified by the German physicist Heinrich Rudolf Hertz (1857–94), briefly mentioned earlier (uncle of Gustav Ludwig Hertz), whose research on Maxwell's discovery of electromagnetism led to his construction of an experimental machine that could produce and detect radio waves in the very high-frequency (VHF) and ultra-high frequency

(UHF) bands. Heinrich Hertz's work was the first successful proof of the existence of electromagnetic waves. It ushered in the industrial research and development of radio, radio astronomy, television, radar, satellite communications, and mobile phones, which are now very much part of our daily lives. The international unit of frequency of one cycle per second, or hertz (Hz), was so named in his honor.

THEY THOUGHT HE WAS A MADMAN

Nikola Tesla was born exactly straddling midnight between July 9 and July 10 of 1856 to the Reverend Milutin Tesla and his wife, Duka, in Lika, Croatia. As a young child, Tesla displayed a highly fertile imagination and was a prolific inventor of toy machines. At only five years old, he built a waterwheel that turned steadily in the current even though it had no paddles. His inventions appeared as real as they could be in his mind's eye, and he was able to build them without any need at all for drawings or models to determine how they would work.

It was a physics professor in Karlovac, Croatia, who motivated the young Tesla, who excelled in mathematics, to focus on electricity by challenging him with a statement that it was impossible to build a motor powered by alternating current (AC) rather than by direct current (DC). A desire to perform experiments and to delve into scientific investigations grew in his heart. Tesla enrolled at the Austrian Polytechnic School, which he and members of his family could no longer afford by his third and final year of what was normally a four-year course. He left school to seek employment and his formal education in mathematics, physics, and mechanics was left incomplete.

Tesla's achievements belie his lack of formal schooling. While inventions whirled about in his head, he worked his way from the drafting office to the engineering department of the Central

Telegraph Office of the Hungarian government in Budapest and later, through family contacts, got a job at the Edison Telephone subsidiary in Paris. He finally entered the United States in 1884, in possession of a letter of glowing recommendations, addressed to Thomas Alva Edison himself, from Charles Batchelor, the Paris plant manager who was a close friend and former assistant of Edison's during the time when the latter was working on Alexander Graham Bell's first telephone.

Thomas Edison (1847–1931) was an American businessman and a successful prolific inventor who developed the first practical incandescent light bulb. In the summer of 1884, Tesla presented himself to a harried, thinly spread Edison, who was beleaguered by many challenges from his burgeoning electricity, telephone, and telegraph businesses and by DC electricity, his pet project. Edison had just promised to send a non-existent engineer to solve the power problem of the ship the SS *Oregon*, and, lo and behold, right before him stood an electrical engineer from the Edison Telephone subsidiary in Paris handing him a letter from a trusted friend and colleague. Tesla started the job of repairing the ship's power plant that day and was finished by dawn. His new employer was very much impressed.

Even though he had proved useful to Edison, Tesla quit after one year because Edison reneged on a mutually agreed-on pay for a job he'd finished. He struck out on his own, helped by some big financial backers. That broken promise from Edison might not have been the main reason for his resignation, however. Tesla had been excitedly working out in his mind the details of AC electricity since his days in Budapest and Paris. Edison was financially and emotionally committed to DC electricity, and he would not hear of AC from a mere employee, no matter how gifted. Besides, he did not see how the young employee could make AC work when other scientists before him had only met failure. But Tesla intuitively knew that it would work and that it would be even better in several aspects than the DC version that

Edison was married to. He had known this in his heart as a budding engineer in Croatia inappropriately pitted against a closed-minded professor.

What had preoccupied Tesla during his employment before his immigration to the United States was the rotating magnetic field. We discussed earlier how the principle of alternating current first revealed itself to Faraday, who understood it at the time. The idea waited for Tesla to perfect it in his mind. Like a bolt of lightning, the clear image of a fully developed motor struck him as he was walking in a Budapest park with his faithful friend Anital Szigety, who was alarmed when Tesla stuck his arm up in the air as if in a trance, and would not move.

At this juncture, Tesla had just recovered from a nervous breakdown. Szigety, an athlete and a master mechanic who was credited with prescribing exercises to help his friend regain his health, feared that he might have gone mad. Tesla asked for a stick, and it was only then that he broke his strange posture and proceeded to write on the sand his perfect idea of a motor that would produce a continuous source of electricity.

Similar in principle to Faraday's electromagnetic induction, Tesla's motor differed in that it had two or more alternating currents created by multiple coils which were attached to an electrical source and activated sequentially but out of step with one another (i.e., they were "polyphase"). This set up created a continuous magnetic vortex: a rotating magnetic field that became the heart of his induction motor and of the polyphase alternating current. The continuous current could be transmitted over long distances without a lot of the voltage loss associated with DC electricity. The rotating magnetic field made the process so simple that the motor practically had only one moving part that was subject to wear and tear, minimizing the chance of mechanical breakdown, a major problem that plagued the more complicated DC motors which had many moving parts.

Today, AC current is what flows from the wall socket into our ordinary household appliances. Digital electronics, television sets, and computers, which require the use of DC for continuous and steady voltage flow, are all equipped with a "rectifier" to convert the AC electricity into DC. Mobile phones, laptops, tablets, cameras, and many other portable electronic items use direct current supplied by batteries.

Tesla invented a transformer using the polyphase alternating current system known as the Tesla coil, which could produce exceedingly high voltages. He used this coil to construct impressive lighting, bright sparks, and loud crackles that he used as props to entertain his audiences of friends and the curious. Tesla also lit bulbs wirelessly with energy gathered from the atmosphere and shaped fluorescent and neon lamps into lighted letters. Just as he did in his boyhood, he built boats, this time with a first-ever wireless remote control.

When the tremendously successful American entrepreneur and engineer George Westinghouse (1846–1914) landed the contract to light the 1893 World's Columbian Exposition (popularly known as the World's Fair) in Chicago, he used Tesla's alternating current polyphase system to do so. This facilitated the acceptance of AC as the electricity of choice in the United States and all over the developed world.

"Someday I will go to America and harness Niagara Falls for power," Tesla once said to an uncle after seeing exhibits of waterwheels. He kept his eye on that dream and made it come to life in a big way. The massive and most impressive Niagara Falls Power Plant is the crowning glory of Tesla's major life-changing achievements. Built in collaboration with Westinghouse in 1895, the world's first large-scale hydroelectric power plant stands as a humongous monument to the exceptional genius of Nikola Tesla's polyphase alternating current electricity and a solid testament to the power of a dream.

10
Entanglement

What we call chance is nothing but the sudden revelation of relationships between things.
—Octavio Paz

In his song mentioned earlier in the book, Frank Sinatra has the world dangling from a string wrapped around his fingers while he is sitting on a rainbow. He has magic at his fingertips. From his perch on top of a beautiful, multicolored, radiant crescent of light frequencies, he can make the rain go away with just the twist of a finger and a song. This sounds weird, but impossible it is not. It might even be kooky and wacky, even mysterious, but it is not impossible.

Sinatra's finger is intertwined with the sun and the rain through the vibrations of his song in the same way that the biblical walls of Jericho came crumbling down during the methodical shouting and marching of the Israelite soldiers while blowing the *shofar* (a ram's horn trumpet) and likewise in the way healing is effected by chanting. This has to do with the fundamentals common to them all—vibration, frequency, harmonics, and resonance—and with the rhythmic meshing of waves of probability of everything great and small, seen and unseen, thus unveiling chances, coincidences, and unlikely events as harmonic components of an orchestrated whole, not as isolated and random phenomena.

In the baffling, weird world of the quantum, when two or more particles have previously been together and then separated spatially from each other, a change in one simultaneously changes the other in the same or in a correlated way. This happens even if there is no obvious "local" connection between the two particles and no other known method of communication exists between them. The uncertainty principle of Heisenberg, which is central to the Copenhagen interpretation of quantum mechanics, allows the appearance of such instantaneous connectivity between separated particles. These suppositions were seriously debated and argued against by the giants of physics during the 1927 Solvay Conference in Brussels, and the excitement moved quantum theorists to get busy. Up to the time of vital theoretical physics experiments on connectedness in the years following the 1927 Solvay, physics was predicated on locality, the then-accepted doctrine that an object is affected by its environment only if it has a way of communicating, directly or indirectly, with that environment. For many members of the scientific community, quantum entanglement was just too preposterous an idea to swallow.

That was then, at the height of the quantum's bewildering and mystical days. Since that time, though, scientists have proven that everything is a composite of waves of vibration. All the time, the whole of ourselves, from our quarks to every strand of our hair, hums in harmonic oscillations. At any time, a dominant harmonic structure exists among all frequencies constituting our whole being, and that dominant tune resonates with our environment. Any change in our harmonics is instantaneously reflected in all the vibrational waves we are resonantly tied in with—and that means everything and anything, in different degrees of resonance and connectedness.

For instance, suppose you are thinking of your sister, and you are dearly missing the sound of her voice. Your vibration is strong and attracts her basic frequency at this specific time, making both of you more resonant with each other than usual, so she misses you, too, and she calls you on the phone. People call this "mental telepathy," "mind reading," "being on the same wavelength," or having "the same vibes." The very words used are truly descriptive of the actual process of frequency resonance that gives rise to the fascinating experienced phenomenon.

Rosenblum and Kuttner said that "in principle, any objects that have ever interacted are forever entangled, and therefore what happens to one influences the other. Experiments have now demonstrated such influences extending over more than 100 kilometers. Quantum theory has this connectedness extending over the entire universe."

In 1982, the French physicist Alain Aspect and his colleagues published the results of their research on photons that had been twinned from a single photon, each twin having exactly the same trait as the other. Dispersed to different destinations, the twin photons were shown to maintain their instantaneous linkage with one another despite their great spatial separation. The team noted entanglement across a distance of around eleven kilometers (about 6.8 miles) between several pairs of twinned photons. A stimulus applied to one of the pair was instantaneously echoed by the other. In separate studies, a three-generation set of "mother," "daughter," and "granddaughter" photons were demonstrated to be entangled in energy and time, and two photons that had never coexisted in time before, clearly displayed entanglement, all because they were previously separately entangled with one other photon.

Quantum entanglement may explain the stability of the DNA double helix. DNA proteins, superposed within the double helix, form harmonic oscillators that are of the same wavelength as the DNA, their waves traveling in opposite directions and forming fixed standing waves that go nowhere.

Instantaneous and harmonic connectedness is not confined to the quantum realm. Such connectedness was observed in 1995 in aggregates of matter called Bose-Einstein condensates, seventy years after the Indian physicist Satyendra Bose (not to be confused with Jagadish Chandra Bose, mentioned earlier) and Einstein postulated that bosons (particles with whole-number spin) in super-cooled gas consolidate or precipitate as gross matter. The Bose-Einstein condensate of atoms was found to have a wave-like property spread throughout that displayed patterns of interference and vortex lattices. The precipitate also exhibited the ability to instantaneously transfer information in a non-local and entangled manner, behaving as if it was in the ever-so-tiny quantum state. If this interconnected behavior of these condensates had been found early enough, Einstein would have had the supreme pleasure of having his own work quell his disquiet over his much-despised spooky action at a distance.

On a bigger scale, scientists from the University of Oxford, the National Research Council of Canada, and the National University of Singapore have found two truly macroscopic diamonds—three millimeters in size separated by a distance of fifteen centimeters—to exhibit vibrational entanglement at room temperature. The scientists split pulses of light and shone each split beam separately on each of the diamonds. The light induced the diamond crystal lattices to vibrate, creating a quantum of sound vibration called a phonon, which was then ejected out of the crystal lattice by another pulse of light. They were able to show with 87 percent probability that the pair of diamonds shared one phonon, regardless of their wide physical separation.

The state of connectivity and unity of all things and happenings in the universe is not a new discovery by the physicists and scientists of the modern age. The old Eastern mystics, sages, and indigenous cultures preached and lived such connectivity, and their successors still do. Unity is the central theme woven through their traditions, oral teachings, ancient sacred texts, and esoteric practices. Common terms such as "whole," "complete," "all-embracing," and "all-encompassing" refer to the same thing: oneness. It is "All in THE ALL" and "THE ALL in All," "the One in Many and the Many in One," and is what is meant in "At that day you shall know, that I am in my Father, and you in me, and I in you" (John 14:20, KJV).

TOUCHING SOMEONE WITHOUT REACHING OUT

Despite having experienced the effect of electricity when he got the electric spark from his kite in that thunderstorm experiment, Franklin did not believe in the presence of an electromagnetic force in the human body. He was serving as the American ambassador to France when he was appointed to chair a 1784 French Royal Commission charged by King Louis XVI to investigate a physician, Franz Anton Mesmer (1734–1815), an Austrian visionary who believed in a mysterious idiosyncrasy that he called "animal magnetism," a biological characteristic known today as bioelectromagnetism. It is attributed to the flow of electromagnetic current in living tissues, regularly recorded in electrocardiograms (ECGs), electroencephalograms (EEGs), and electromyographs (EMGs). Bioelectromagnetism is the principle behind the use of TENS units (transcutaneous electric nerve stimulation) and pulsed radiofrequency for pain, "treating fields" (low-intensity, moderate-frequency alternating current electric fields) for treating cancer, electric cardioversion for cardiac arrest and heart rhythm abnormalities, transcranial magnetic stimulation for depression, intense pulsed light for skin conditions, the set of techniques known as Kirlian photography,[9] and many more. It is a phenomenon exhibited by insects, electric eels, bacteria, birds, bats and other animals, and by plants in their photosynthesis and other functions, such as the ingenious way used by the Venus flytrap in catching its food.

The French Royal Commission that Franklin led did not investigate whether or not mesmerism effected relief or cure. It contrived to ask a question that could not be answered at that

---

[9] Kirlian photography, named for the Soviet inventor Semyon Kirlian, captures objects' electrical coronal discharges; some proponents of the technique believe that such a discharge represents an aura around the object.

point in scientific history: whether evidence existed for Mesmer's animal magnetism. Not having the current knowledge regarding bioelectromagnetism, the commission concluded that Mesmer's animal magnetism lacked scientific evidence, and he was censured.

Mesmer studied medicine in Vienna, and his doctoral thesis was titled "On the Influence of the Planets on the Human Body" ("De Planetarum Influxu in Corpus Humanum"), a dissertation on the effect of tides created in the human body by the movement of the sun and the moon and other celestial bodies, thus influencing human health and disease. He believed that everything was linked by a subtle unseen fluid that filled the whole universe, that every living thing possessed this substance in a balanced manner, and that some people (including him) were endowed with more of it than others. He thought they could utilize the substance at will, enabling them to magnetize and bring about healing in human beings and animals.

Mesmer strove to augment this innate gift and fine-tune his magnetizing ability by adopting some of the techniques of Johann Joseph Gassner, a much-respected Catholic priest and well-known exorcist and healer in Vienna. Mesmer always maintained that even though the power of animal magnetism was strong in him, it was much stronger still in Gassner. Interestingly enough, the priest's popularity waned after Mesmer pointed out that it was Gassner's innate animal magnetism, and not the phenomenon of religious exorcism, that was behind the success of his method.

The early practice of magnetism or mesmerism, the eighteenth-century predecessor of hypnotism developed by Mesmer and his disciple the Marquis de Puysegur, involved "tuning in" with the patient and establishing rapport with him or her at the very start of a session. This forged a bond between the magnetizer and the patient that allowed the practitioner to control the "healing crisis" symptoms. Those symptoms included seizures, fever, limb movements, unusual body feelings, and other effects that were elicited in the patient by the magnetizer. The crisis was a necessary mechanism to start the healing process, although many weaker manifestations were preferred to a few strong ones, both by the patient and the practitioner.

Victor Race, a young man from a family who had served the Puysegurs for generations and thus was closely interwoven with various members of the Puysegur family, became one of the marquis's earliest patients. The magnetizer noticed a strangely different crisis in him. He would sing out loud the very song that Puysegur was silently singing in his head. He would fall into a trance-like sleep in which he was described to be more aware and more awake than his normal state of wakefulness.

Adding to the strangeness of this account of connectedness is the story of Mesmer's pet canary. This bird lived in an open cage, and every morning it would fly to its sleeping master's head and sing him a song to wake him up. It would then keep him company at breakfast. Mesmer was said to be able to instantly put the bird to sleep or awaken it with just a stroke of his hand. In the morning of Mesmer's passing in his sleep, the canary did not sing or get out of its cage. It stopped eating and died shortly thereafter.

Mesmer also successfully utilized techniques and props—such as music, magnetized water into which subjects dipped their body parts, and magnetized trees, which the subjects touched or were attached to by cables—in order to heighten the emotional state of the subject or subjects in a group setting, himself included. Today, his method is used in the fields of entertainment and in medical-psychological methods of treating mental and emotional illness.

Mesmer gained many followers, physicians and lay people alike, including members of the European nobility such as Empress Maria Theresa and Wolfgang Amadeus Mozart, who practiced mesmerism to alleviate and cure various ailments. He also attracted considerable negative attention from the medical community in Vienna as well as in Paris, where he relocated following ostracism by his Viennese colleagues. After the Royal Commission's investigation, mesmerism was shunned once again by the medical community. That rejection, and the start of the French Revolution, occasioned its decline in popularity. Staunch practitioners, though, continued the movement in secret, and it thrived in the underground.

To "mesmerize," to hold in spellbound attention, has its roots in Mesmer, and the practice of hypnosis hailed from mesmerism. The term "hypnosis" comes from the Greek word *hypnos*, meaning sleep, after the god of sleep, Hypnos, and from the trance-like state that strongly resembles somnambulistic sleep established on mesmerized subjects.

THE BEGINNING OF TREE HUGGING

Grover Cleveland "Cleve" Backster Jr. (1924–2013) established the polygraph section of the US Central Intelligence Agency (CIA) after World War II. He was the CIA interrogator and lie-detector expert until 1960, when he founded his own polygraph school in New York City, training and certifying members of the police force, the Federal Bureau of Investigations (FBI), and the CIA in interrogative polygraphy. He later moved to San Diego, where he opened the Backster School of Lie Detection. It was when he was in New York that he had a life-changing experience while watering a dragon tree plant in his office.

Familiar with the work of Jagadish Chandra Bose on the electric potential of plant-cell membranes, it occurred to him that he might be able to measure the time it took for water to travel from the root to the leaves with his lie-detector polygraph machine. Lie detection works by attaching electrodes on three fingers and around the chest of the subject. The electrodes are hooked to a machine with multiple needles to record the flow of electrical current on moving paper.

The machine measures and records the client's response to specific questions, based on the premise that if a person is lying, there is an increase in the skin's electrical conductivity, heart rate, blood pressure, and rate of breathing, all of which present as an increase in spiking activity of the machine's needles and height of recorded spikes. All the responses are due to sympathetic nervous system excitation, a state of nervousness when someone is aroused, scared, startled, suddenly paying close attention, or having a mental picture of something that makes the subject anxious or filled with emotion. Especially evident on the palms and soles, this sympathetic response causes an increase in sweat secretion, which boosts the skin's ability to conduct electricity.

Backster proceeded to clamp electrodes to both sides of a leaf of the dragon tree and started his recordings. What happened next would forever change the course of his life and the way some members of the human race would view all life forms on earth. He noticed that the needles displayed tracings that were similar to those recorded when subjects have a mild emotional reaction to interrogation. It was not at all the enhanced electrical conductance that he expected as a result of increased hydration of the plant.

He then thought that since the dragon tree was behaving like a person during interrogation, he would utilize people-oriented techniques for eliciting strong responses. He decided to use the "threat to well-being" technique, so he dipped a leaf of the dragon tree in hot coffee. He did not get a response. He then thought of something more drastic and decided he would burn the leaf instead. The needles wildly spiked suddenly, even before he could get the matches. The plant seemed to be affected by his thought of burning it! The needles jumped again the moment he came into the room with matches in hand, which to him indicated a reaction to his intention to carry out the threat. When he pretended to go about burning it without really meaning to, the plant did nothing. The plant behaved as if it could read his mind. His plants also seemed to give a pattern specific to the death of living tissue and also reacted to his and other people's emotional states, even when separated by several miles.

On one occasion, when a visiting scientist from Canada came to observe his plants' reactions, the plants remained quiet, and the graph lines stayed flat. Intrigued, Backster asked the visitor what he did to plants in his experiments. The physiologist told him that in order to get the dry weight of the plants for analysis, he roasted them in the oven. Backster suspected that his plants had gone into a fainting spell to protect themselves. They recovered forty-five minutes after the visitor had left for his flight out of New York; making them presumably feeling finally safer.

They seemed to have extrasensory perception (ESP). Backster did not wish to describe the result of his experiment in this manner, however, because ESP connotes impressions beyond the five senses that humans possess and plants are not supposed to have. He chose "primary perception" instead, because the term describes something intrinsic and basic to the nature of the plants.

Cleve Backster published his experimental data in a paper titled "Primary Perception," in a 1968 issue of the *International Journal of Parapsychology*. A lot of lay people were intrigued and excited about this "Backster effect," and his article was read widely in the United States and abroad. People started to have gardens and began talking to their plants and loving them. Nevertheless, the scientific community labeled him a crackpot and minimized his work as a mere product of various environmental factors in the laboratory, such as static electricity buildup, humidity, and other miscellaneous determinants.

Some critics attributed his observations to his own personal telekinetic abilities, reminiscent of the charges hurled at Dr. Franz Anton Mesmer regarding his early form of hypnosis two hundred years earlier. A panel of biologists at the 141st annual congress of the American Association for the Advancement of Science declared his findings unsupportable.

The failure of other laboratories to replicate his results is hardly surprising when seen in the light of quantum entanglement and the observer effect. Backster and his live experimental objects were very much attuned to one another, and resonant harmonization of frequencies would be expected to be strong between them, thus inducing the collapse of Schrödinger's wave function and giving rise to the Backster (observer) effect. Plants are composed of cells with membranes, and all cells with membranes have microtubules. Microtubules have been shown to be integral to electromagnetic quantum biological processes that have been found to resonantly interact with the cells' surrounding electromagnetic environment, as has been discussed in previous chapters.

It is logical, then, to postulate that each and every mechanism of a plant's response to its environment—such as the utilization of light, temperature, gravity, touch, and volatile chemicals; the communication between plants through underground plant-root fungal network; the identification of kin, community, enemies, growth enhancers, and pollinators—is accomplished by resonant electromagnetic information reception and processing, just like what goes on in the microtubular structures of human tissue cells.

Reckoning on the still-unfolding story of a plant's life alone, perhaps the scientific community should refrain from labeling observations that do not conform to the prevailing scientific mold as pseudoscience, much less rejecting them outright because of the present inability of some of its membership to see beyond data obtained from contemporary experimental procedures using available tools of our time or because of personal or herd bias or agenda. History has shown that scientific paradigms change as more information is acquired using more advanced tools and methods, and as greater understanding is attained based on lessons learned from research and experience. The least that scientists and honest skeptics can strive to do in order to abide by the truth is to keep an open mind; one must continue to ask questions, dig deeper into troubling patterns, and adopt a wait-and-see attitude.

Backster made no references to quantum phenomena to explain his observations. He only knew deep in his heart that he had rediscovered something that had been known by the ancients and by modern sages. He wrote, "We keep hearing about non-time-consuming communication from Eastern sources.... They tell us that the universe is in balance; if it happens to go out of balance someplace, you can't wait ... for the imbalance to be detected and corrected. This non-time-consuming communication, this oneness among all living things, could be the answer."

Quantum entanglement. This inconceivable state of integrated oneness of everything with everything else—including us, human beings—is inextricable from the endless dynamic dance of being. It is permanent and is forever maintained over space and time and across time. And Frank Sinatra may still be sitting on a rainbow making the rain stop and the golden sun come out in my own private world with the twirl of a string of music around his magical fingers.

11

The Gravity of Einstein's Relativity

Spacetime tells matter how to move; matter tells spacetime how to curve.
—*John A. Wheeler*

The world Newton grew up in was a place where mass, space, and time were fixed and constant, and the universe had a comfortable permanence and reliability. The universe's inhabitants of stars, planets, space, and their contents were set in specific geographic territories and followed rigid laws regarding their motion and spatial relationship with one another. His law of gravity stated that, without exception, all objects in the universe attracted one another with the force of gravity. The distances between celestial bodies were simply dictated by the strength of the gravitational attraction between their differing amounts of mass. It was that straightforward and simple.

Newton formulated his first law of motion based on Galileo's concept of inertia, which denotes the tendency of an object to resist any change in its state of rest or motion. The law states that "every body continues in its state of rest, or in uniform motion in a right [straight] line, unless it is compelled to change that state by forces impressed upon it." A ball resting on a flat surface will stay there and not move until an action on the ball creates an imbalance of the total forces that are exerting influence on the ball, which makes it move. By the same token, if the ball is rolling on a flat and smooth surface at a steady clip without friction, then it will just keep on rolling at constant speed unless something happens to create an imbalance of forces.

When imbalance is induced by some force applied to it, Newton's second law of motion comes into play. The acceleration of an object as produced by a net force is directly proportional to the magnitude of the net force, in the same direction as the net force, and inversely proportional to the mass of the object. Force is the product of mass multiplied by acceleration, as expressed in the following equation: $a = F/m$, or $F = ma$.

Newton's third law of motion tells us that for every action, there is an equal and opposite reaction. This principle is the reason why doors swing back and forth and rubber bands recoil and how powerful propulsion rockets thrust forward.

Newton's notion of inertial motion extended Galileo's concept to include all objects, and then he proceeded to attribute this inertial motion to the mass of the object. The greater the mass, the more inertia it contains and the greater the force that needs to be exerted in order to produce a change in its state of motion or state of rest. A cannonball rolls a much shorter distance than a marble on the same surface, since the cannonball has greater inertia because of its greater mass.

Mass is equivalent to weight in kilograms (or pounds or whatever measurement is used) only on earth. In deep space, far away from any gravitational influence, weight drops down to zero. A sixty-kilogram person weighs only ten kilograms on the moon, where gravity is one-sixth that of the earth, but the person's mass remains sixty kilograms because the numbers of protons, neutrons, and electrons that make up the mass do not change. According to Newton, every motion that occurs in space is measured by time: an unchanging and enduring mathematical reality that cannot be perceived. He said:

> Absolute, true, and mathematical time, in and of itself
> and of its own nature, without reference to anything

external, flows uniformly and by another name is called duration. Relative, apparent, and common time is any sensible and external measure (precise or imprecise) of duration by means of motion; such a measure—for example, an hour, a day, a month, a year—is commonly used instead of true time.

And regarding space, he wrote: "Absolute space in its own nature, without relation to anything external, remains always similar and immovable."

Gottfried Wilhelm Leibniz (1646–1716) was a prominent German mathematician and a contemporary of Newton's. A philosopher who greatly influenced the great thinker Immanuel Kant (who went on to influence Western thinking in many ways), Leibniz rejected Newton's idea of absolute space. To him, space did not have a separate and independent existence, with objects inside it located at a certain distance from a specific side of the absolute space. That notion would require space to have specific limitations in shape and size, for example, a square box just so high, so wide, and so deep, which he could not accept. He thought that it is the geometric arrangement of objects that dictates the distances between them and determines their positions relative to one another. Geometric relationship is what defines space.

But what truly bothered scientists was their discovery, little by little, of subtle inconsistencies in the theory of absolute space, which hindered their ability to explain certain phenomena of nature. Einstein saw a need to rethink these absolutes and in due time came up with his two relativity theories.

THE PRINCIPLE OF RELATIVITY

In any scientific inquiry into nature, all natural laws are assumed to apply in the same manner every time, and not to change from

167

one person to the next. The principle of relativity specifies that the natural physical laws should appear the same in every inertial frame of reference, where the observer is at rest or is moving uniformly. However, different relative perspectives may be seen from non-inertial frames of reference, where the observer is accelerating, slowing down, or moving irregularly.

This kind of relativity that we commonly experience is what Galileo described in his popular work *Dialogue Concerning the Two Chief World Systems: Ptolemaic and Copernican*, which brought about a charge of heresy by the Roman Catholic Church. At some point, Galileo invited his readers to pretend that they were in an enclosed cabin on a ship sailing smoothly and uniformly, with no way to look outside. Inside, things appeared to be as if the ship were docked and very still, where insects flew about normally, their speed not varying, whether they were going toward the bow, side, or stern of the ship. The fish behaved the exact same way, swimming normally in their bowl of water with no change in their effort, whether they were going toward the bow or the stern of the ship. If you were there and you jumped high in place, you would land on the same spot rather than on another spot not far from it, on the assumption that the floor would have moved out from under you while you were in midair.

If you jumped toward the stern of the ship, you would not cover any more floor space than you would if you had jumped toward the bow of the ship, even though the ship was moving forward while you were jumping. You would think that the floor of the ship would move away while you were airborne and would contribute to the total floor distance covered by your jump. Likewise, if a friend threw a ball from the stern to you at the bow, then she would not need to throw it harder to you than you would need to throw it to her, just as if you were on the non-moving ground. You would think that she would have to, because your position would be moving forward and farther away from the ball while it was in flight towards you.

The same scenario is true today inside a jumbo jet flying steadily at hundreds of kilometers per hour, high up in the sky. Passengers are unaware of their forward motion, and everything seems as if the airplane is on the ground. They are in an inertial frame of reference. On looking out the window, a passenger may realize that he is moving—in a non-inertial reference frame— relative to cloud formations up close that quickly disappear and to distant mountains that slowly recede out of sight.

The principle of relativity was what Jules Henri Poincaré (1854–1912) a French theoretical physicist, mathematician, engineer, and science philosopher, described in a philosophical paper in the nineteenth century. The strange and uncomfortable scenario was already known then and had puzzled people long before his time. The principle of relativity weighed heavily on Einstein and he, too, devoted time to the idea and its relation to the finite speed of light in a vacuum, which had been discovered more than two centuries earlier.

THE CHALLENGE

Newton's laws were consistent, predictable, and commonly observed in daily life. They were widely accepted, and they ruled the landscape of everyone's reality for a long time. In 1905, a year of creative surge for the twenty-six-year-old patent office "slave," as Einstein described himself, the classical Newtonian notions of a consistent and unchanging universe and unbending laws of time, gravity, and motion were shattered. The end of the laws of absolutes was long in coming, but arrive it did, facilitated by the many advances in scientific physical theories on nature and reality and scientists' understanding thereof.

In 1898, Poincaré outlined a new method of synchronizing clocks using light signals, on the premise that light travels at a steady speed. Shortly before that, in 1873, James Clerk Maxwell had unified four equations on electromagnetic fields in what are now known as Maxwell's equations. Furthermore, Lorentz had already derived from Maxwell's equations the change in frequency of waves of electromagnetism traveling between inertial frames of reference. As the observed object in an inertial frame moves farther away from an observer's inertial frame, the frequency of electromagnetic waves from the observed object gets lower and its wavelength becomes longer. This frequency change later came to be known as the Lorentz transformations.[10]

Maxwell had shown that light is a propagating electromagnetic wave and that all propagating waves travel at the speed of light (written as c), which is 186,282.4 miles (299,792.4 kilometers) per second in a vacuum. For all practical purposes, this speed is pegged at 186,000 miles or 300,000 kilometers per second. No wave is able to travel faster than that. The new theory of light as a wave of electromagnetism required the presence of a light-bearing or "luminiferous" ether that occupies space as a medium for wave propagation. Newton had suggested the existence of an oscillating ether medium in his *Third Book of Opticks* in 1718. Every single thing in the universe would move in this unseen omnipresent medium, which would offer no resistance to the movement of objects within it.

In 1887, Edward Morley, an American chemist, and Nobel laureate Albert Michelson, an American physicist, set about to measure the speed of the earth through the ether, based on the known speed of light traveling as a wave. They thought that the speed of a moving object such as the earth should be measurable

---

[10] Offered by Lorentz before Einstein's special relativity, the Lorentz transformation equations mathematically express the concepts of special relativity: that there is no absolute space or absolute time and that the speed of light in a vacuum is constant and independent of the motion of the observer or the source of light. The Lorentz transformations show how observers moving at different velocities may measure different distances, elapsed times, and sequences of events.

using the speed of light and the ether as reference points. The presence of ether flow would occasion a change in light pattern, as detected by an interferometer to which light was reflected by mirror-beam splitters.

What they found surprised them. Contrary to expectations, they were unable to measure the speed of the earth at all: it was as if the earth stayed in place, even as it whirled around the sun at 18.75 miles (30 kilometers) per second. At the same time, they found the speed of light to be equal in all directions. They could not find an ether for the earth to move relative to, although the experiment itself did not entirely rule out the existence of such a substance. Still, scientists interpreted this finding as proof that the ether did not exist at all.

IT WAS A VERY GOOD YEAR

In the miracle year of 1905, when Einstein was only twenty-six years old, he obtained his doctorate in physics from the University of Zurich and published his theory on the photoelectric effect that would go on to earn him the 1921 Nobel Prize. In that sensational year, he churned out four earthshaking, belief-shattering scientific papers. The first article, which contained his groundbreaking photoelectric effect that was discussed at length earlier, was submitted to the journal *Annalen der Physik* (*Annals of Physics*), and was titled "On a Heuristic Viewpoint Concerning the Production and Transformation of Light."

His second paper, "On the Movement of Small Particles Suspended in Stationary Liquids Required by the Molecular-Kinetic Theory of Heat," was on Brownian motion, which refers to the seemingly chaotic and random movement of tiny pollen particles suspended in liquid. The third article that Einstein submitted was titled "On the Electrodynamics of Moving Bodies," which became the "theory of relativity" when Max Planck dubbed it thus in reference to the theory's basis on the principle of relativity. We know it now as Einstein's special theory of relativity.

The fourth work, "Does the Inertia of a Body Depend Upon Its Energy Content?" was published in November 1905 and officially introduced the equivalence of mass and energy; it also revealed for the very first time the most famous and humanity's singularly significant equation: $E = mc^2$. Poincaré had published a similar equation, $m = E/c^2$, in 1900, but his treatment of the subject was not as thorough and as detailed as Einstein's. This last paper laid bare scientists' raw realization that mass and energy are equivalent relative to the speed of light.

Because of the constancy of the speed of light, mass and energy can be converted into each other according to this equation, which tells us that mass and energy are the same. Nuclear fission, the splitting of the nucleus of an atom, would bring about enormous energy radiation relative to the mass of the element whose atomic nucleus was split.

Einstein became a professor at the University of Zurich, and then at the Berlin Academy of Sciences. He was often invited to the United States as a visiting professor, and he was teaching there when Adolf Hitler and the Nazi party came into power in Germany. On his way home to Berlin after his professorship duties were complete, he learned that his house had been raided by the Nazis. Being of Jewish descent and opposed to war in principle, he deemed it best to renounce his German citizenship and to remain a Swiss citizen.

Subsequently, he emigrated to the United States became a citizen in 1940, although he still retained his Swiss citizenship. He took a post at the Princeton Institute for Advanced Study (close to the prestigious Princeton University where he where gave lectures) and remained there until his death in 1955.

SPECIAL RELATIVITY

In that productive year of 1905, Einstein was still at the Swiss patent office in Bern, where he had worked for three years to support his wife and their young son. The Swiss patent office was located on Kramgrasse (Swiss-German for market alley), the city's center of commerce. It was and still is an expansive boulevard with covered arcades, lined with market stands that have now been supplanted by high-end stores. The street is decorated with three fountains and sculptures and an especially famous landmark clock tower, the Zytgloggeturm, which features a two-faced astronomical clock with many interesting adornments, but the most charming wonder of all that people eagerly wait for are the chimes. Starting at four minutes before the hour, the chimes start, and decorative figures of a jester, bear cubs, and a rooster dance to their music. The performance is followed by the striking of the gilded great bell in the cupola by a gilded wooden figure in full armor to announce the arrival of the hour. The captivating show is brought to an end by the crowing of the rooster.

The chiming two-faced clocks and their characters were only a small part of the motley scene on Einstein's Kramgrasse. There were people busily shopping or scurrying to the train station, getting off and on train cars, and rushing on and off the platform; loud bells, clanging rails, and tooting train whistles; train lights, stop lights, and arriving and departing cars, carriages, and streetcars; and luggage piled and scattered about. The interplay of light, sound, and motion must have made a sharp and powerful impression, whether consciously or subconsciously, on the bright young physicist.

As a child, Einstein had wondered what it would be like if he could ride alongside a beam of light. By 1905, he was ready to reach beyond his boyhood musings to postulate that the speed of light is the same for all observers, regardless of their motion relative to the light source. So, to two observers at two different velocities, the speed of light would be the same as if they were at rest or non-accelerating. Electromagnetic light traveling as a wave sets the universal speed limit for all propagating waves and is the maximum possible velocity any object can reach. If light is slowed down as it shines through matter, for example through water, then other objects also traveling in the same medium will be slowed down correspondingly.

According to the then-accepted Newtonian principles, objects move from one absolute space to another absolute space without regard to other objects. This is called absolute or inertial motion. An object just moves, independent of and unaffected by anything around it. Although Newton's laws of absolutes apply at slower speeds, they lose applicability when speeds are closer to the speed of light—one of the inconsistencies that bothered Einstein. He sought to establish inertial motion as relative and that there is no such thing as absolute rest in the inertial frame of reference, counter to the prevailing Newtonian idea. His new concept dealt only with inertial motion and excluded accelerated or non-inertial motion: the reason for the term "special relativity."

Since absolute motion is the movement of an object from an absolute space to another within Newton's absolute time, the new special relativity equations also effectively eliminated independent space and independent time. To Einstein, time, motion, and events were dependent on the position of the observer relative to a reference point. One person's experience of two simultaneous events might actually be two sequential events when viewed from the perspective of a second observer somewhere else.

Einstein discovered that three-dimensional space is woven together seamlessly, with time as the fourth dimension of what is now called space-time that permeates the universe with an unbroken and boundless geometric and mathematical framework. Each dimension is dynamically linked: nature's objects move and their movement and events flow relative to one another.

Four-dimensional space-time was not Newton's independent, static, and absolute space and time. Based on Einstein's new special relativity theory, Newton's classical world, where time ticked at a steadfast speed, had to be given up.

The equations of special relativity did not need the ether to be factored in, and again, the common interpretation was that the ether was not necessary and therefore non-existent, consistent with the Morley-Michelson findings. Special relativity laid out the relationship between energy, mass, and light in the most mortally destructive equation in history: Einstein's $E = mc^2$. A few decades later, the power of the equivalence of matter and energy would be unleashed against humanity.

Before the Second World War broke out, Einstein warned President Franklin Delano Roosevelt of the possibility of German bomb development based on nuclear fission, prompting the United States to develop the atomic bomb. Einstein did not really wish for this scientific discovery to be used. Ever the pacifist, he discreetly lobbied hard and desperately promoted the creation of an international body to have control over the monster—to no avail. Hiroshima and Nagasaki sadly got buried under radioactive mushroom clouds anyway.

Only approximately 700 milligrams (or 0.02469 ounces) of mass of the 64 kilograms (141 pounds) of enriched uranium was converted into energy equivalent to about 15 kilotons of TNT when the atomic bomb, codenamed "Little Boy," was dropped on Hiroshima. This miniscule amount of matter, less than one-third the mass of a dime, was annihilated through nuclear fission to unleash the ferociously deadly energy that created a gigantic fireball 50 percent hotter than the surface of the sun and forced imperial Japan to surrender nine days later.

The equations of special relativity predict that time undergoes "dilation" for an object moving close to the speed of light, or, in plainer language, time slows down when one is moving very fast.

Antiaging enthusiasts may wish to embark on space exploration traveling at near-light speed in order to experience the effect of time dilation (and play among the stars with Ol' Blue Eyes), which will allow them to age slower than their earthbound counterparts. When they get back to earth, though, their contemporaries and even their twins (in what is known as the "twin paradox") will have aged much more than they, and they will need to adopt younger families and build a younger circle of friends.

The military application of time dilation has had a direct impact on the public. The Global Positioning System (GPS) installed by the US military to the tune of over $10 billion for navigational purposes has twenty-four satellites equipped with atomic-precision clocks to beam information to earthbound receiving units. They are able to transmit time to a precision within fifty-billionths of a second of local time and to provide maps and altitude, latitude, and longitude with accuracy to within fifteen meters. Commercial industry has been quick to apply this new navigation technology for general public consumption. Cars are now equipped with GPS devices, as are airplanes, blimps, ships, boats, trucks, hot air balloons, equipment for bridge-building and oil exploration, smart watches, and, closest to our personal realities, our very own smartphones, let alone Amazon's Alexa or Apple's Siri, either one of which must be making the author of *1984,* George Orwell, squirm in his grave (may he rest in peace).

Special relativity also says that a fast-moving rod is measurably shorter than a slow-moving rod, a phenomenon called length contraction. If a rod travels close to the speed of light, it will become so contracted as to become only a point. Obviously, this scenario is intuitively not very accommodating to the imagination, much less to rendering proof.

Special relativity integrated space, time, mass, and energy, but not gravity. This failure necessitated the eventual development of the theory of general relativity.

The blending of space and time by special relativity made the theory incompatible with Newton's rigid laws of gravity. Those laws allowed no exceptions to the attractive force between two or more objects, since they were based solely on their individual masses and distances from one another. It became imperative to provide an explanation for the mechanism by which gravitational attraction between objects with mass influences their positions and motion in space-time and by which planet earth faithfully revolves around the sun and the moon around the earth.

Einstein worked hard to develop a theory that would solidly reconcile space-time with gravity. Ten years later, he revealed the staggering theory of general relativity, which warped space-time around all gravitationally attracted matter in varied interplays of relationships between position, motion, time, and space—finally in integrated concordance with the net effect of gravity. Einstein had offered the world a fine example of quantum entanglement on a grand scale, whether or not anybody fully knew it.

GENERAL RELATIVITY

The connection between the human body and the earth is accepted to be due to gravity. Gravity gives us a feeling of being securely grounded, assuring us that we are not, at any moment, in danger of falling off the face of the earth or of flying away to the moon.

The attractive force of gravity acting between two bodies is directly proportional to the product of their masses and inversely proportional to the square of the distance between them. This is Newton's law of universal gravitation. The closer the bodies are to each other, the stronger the gravitational attraction between them; the farther apart they are, the weaker the attraction. The law of gravity dictates the positions of bodies according to their gravitational attraction alone. But this situation holds true only for objects reasonably near one another, and not necessarily for those substantially far from a massive object.

The general theory of relativity posits that matter curves the very structure of space-time, and the resultant curved and warped space dictates how objects move in time according to their spatial relation to one another, resulting in an intertwined dynamic relationship between time, space, matter, and energy. Any ripple in the warp in one place causes a change everywhere at exactly the speed of light. This is akin to having two heavy balls of different weights on a trampoline, each one affecting the other's position and each changing the other's motion, depending upon how they individually deform the trampoline's surface.

The notion of the warping of space-time around gravitationally attracted objects explained the hugely problematic discrepancy between Newton's law of gravitation and Mercury's real mapped journey around the sun. Scientists had observed the planet's position closest to the sun (called its perihelion) to be inconstant; they noted that it was moving slowly in a wobble ("precession") around the sun. Calculations based on Newton's laws of gravity, taking into account the effect of the sun and the planets, predicted an amount of precession that was short of what was actually observed. According to Newtonian law, Mercury should trace a predictable route as well as distance from the sun. General relativity theory said otherwise. Due to the proximity of Mercury to the sun, a massive, gravitationally attractive body, the resultant warping of space-time in the sun's gravitational field would affect Mercury's orbital path around the sun. Einstein used the equations of general relativity to predict the precession of Mercury's orbit. His hunch proved to be correct: the planet's positions predicted by the equations exactly matched the actual orbital observations.

Then, in another divine spark of inspiration, Einstein extended general relativity rules to light. He expected the path of light through gravitational fields to follow the warping of space, just as mass does. He used the equations of general relativity to predict the degree of deflection of light as it passes through the sun's gravitational field. His calculations doubled the degree arrived at when using Newton's equations of gravity. The concept was generally not seen as a big deal then because it was difficult to test, and the experiments performed were still based on premises and procedures compatible with the classical laws of Newton.

Serendipitously, a synchrony of natural events gave the blasé scientific community an excellent chance to test Einstein's theory of general relativity. A total solar eclipse was predicted for May 29, 1919, and Sir Frank Watson Dyson, Astronomer Royal of Great Britain, thought of a perfect experiment to do exactly that. Light from the Hyades star cluster (which is normally not visible during the day) would be crossing the sun's gravitational field on its way to the earth at the exact time of the eclipse. The darkness during the total eclipse was expected to allow the star cluster to become visible near the hidden sun, but in a position in the sky different from its location at night. This change would result from the deflection of its light as the cluster passed through the sun's gravitational field, a phenomenon called "gravitational lensing."

An excited group of scientists headed by English astronomer Sir Arthur Eddington took measurements of the positions of the Hyades star cluster in January 1919. Then, on May 29, he sent two expeditions to ensure visibility in at least one location. One team went to the remote island of Principe off the west coast of Africa, which he joined, and another team went to Sobral, Brazil.

In November 1919, Sir Eddington released the team's long-awaited findings. Hyades's position had dutifully shifted at exactly the same degree of deflection as had been plotted according to the predictions of the general relativity equations. Overnight, the news catapulted the then-still-obscure Einstein into the limelight of celebrity, both socially and professionally. His thick, rumpled, and slightly graying hair framing a kindly face graced newspapers and magazines and would continue to do so in the ensuing years. Gravitational lensing has become an important tool in deep space observation by astrophysicists. In 1999, scientists found that the bending of a quasar's single beam of light from 10.4 billion light-years away as a result of the lensing effect as the beam passed through a galaxy aligned in its path, had created four points of light in the form of a cross, dubbed the "Einstein Cross."

In 2010, a UC Berkeley team launched a high-tech inquiry into a distant pulsar, which is a remnant of a supernova explosion and the supernova's white dwarf companion. In the presence of a phenomenally strong gravitational field, as is present around the pulsar, Einstein's general relativity theory predicted that light would slow down and its wavelength would increase, resulting in a gravitational "redshift." The Berkeley team delivered the goods to Einstein by clearly showing the effect of gravity on the flow of time: gravity slowed time down, measuring precisely as general relativity predicted, with an accuracy ratio of one to one hundred million. Now, that is some precision!

The accuracy of the predictions of general relativity theory and its successful application over the years have established general relativity as one of the pillars of today's physics, the other being quantum mechanics. General relativity does not rightly apply to the quantum and vice versa. Science tells us that whenever its equations are extended to the quantum, general relativity laws break down, and gravity ends up in infinity—a black hole. The same holds true for quantum equations. $E = hv = mc^2$ creates more mass as energy gets bigger, so that mass becomes so dense that it implodes on itself and it ends up being a black hole—again.

OF BLACK HOLES AND WORMHOLES

The distinguished French mathematician Pierre-Simon Laplace speculated about its existence; Einstein's general relativity predicted its presence; Karl Schwarzschild found it in his solutions to the field equations of general relativity; and finally, John A. Wheeler coined the term "black hole" that we now know it by.

A "singularity" is a central point out of which everything emanates, according to ancient Hebraic and Eastern religious teachings. Mathematically speaking, the core of a black hole is such a singularity—a finite mass of infinite density, but is infinitesimally tiny that it cannot be measured. A black hole has zero size. It is a central mass so dense, with a gravitational pull so strong that not a single thing can escape its tight grip, ever. Not even light.

The quantum uncertainty principle dictates that a black hole vortex core has some amount of (or non-zero) temperature and entropy; therefore, it should lose energy. This property allows a particular black hole that loses energy and mass faster than it can gather from its accretion sources to dissipate over the course of time. In other words, logic dictates that black holes with net energy and mass loss would eventually evaporate to nothingness. This is said to be particularly fitting for micro–black holes, which should disappear quicker than their bigger counterparts.

This process might be what goes on in particle creation and annihilation in the zero-point field and likewise, in the conception and death of human beings. (As described in the next chapter, the zero point is the theoretical absolute zero temperature of -273.15 degrees Celsius, or -459.67 degrees Fahrenheit.)

The law of energy conservation negates the loss of information and energy in the universe. Even though light cannot be seen from the black hole, information is still retained. Scientists have postulated, and have painted a word picture, that on the other side of a black hole, there is a "white" hole from which energy, light, and information gravitationally sucked in and incorporated into the black hole singularity are eventually released in a massive burst of energy in the same manner as the big bang.

Karl Schwarzschild defined the existence of singularities from Einstein's equations on general relativity, but the concept of particles as singularities occurred to de Broglie as he worked on solutions to Schrödinger's equations related to his theory in which "pilot waves" guide particle singularities in Schrödinger's quantum wave function.

Scientists have proposed a black hole singularity to be the center of a torus into which everything spins in a vortex and meets in perfect symmetry. Much like the perfectly calm eye of a storm vortex, the singularity is completely still. It is the seat of standing waves of vibrations beyond the tiniest measure, the Planck scale.

Scientists have artificially produced vortex cores in the lab. Researchers in Japan induced vortex core formations in nano-size ferromagnetic disks by exposing them to a weak magnetic field. They demonstrated that these vortex cores vibrate in frequency resonances ranging from 200 MHz to 600 MHz and could be magnetically tuned to a wide range of resonant frequencies.

William Thomson, First Baron Kelvin (known as Lord Kelvin), a Scotch-Irish mathematical physicist and engineer, hypothesized that atoms may be stable "vortex rings" (imagine smoke rings) in an ether that pervades space and are the building blocks for the elements in the periodic table. Curious about this property, researchers Dustin Kleckner and William Irvine created knotted vortex loops (imagine knotted Moebius bands) in water in a 2013 study reported in the journal *Nature Physics*. They noticed that the knots changed their shapes all the time: they seemed to stretch and move away in opposite directions and then move toward one another again, reconnecting upon their collision while annihilating some part of one another, changing shapes as they evolved.

The phenomenon appears to be indicative of Heisenberg's uncertainty principle at work in the black hole vortex. It has to do with the frequencies of the waves of probability in the vortex of an object and the principle of resonance. If the dominant frequencies of the composites of probability waves of two vortices are resonant or harmonically matched, the vortices are attracted towards each other and they get entangled. They may stay together as one vortex—but not forever, as displayed by the behavior of the water vortex rings in the above-mentioned experiment.

Those probability waves least resonant with the core singularity of the vortex of combined frequencies may disconnect temporarily and then rejoin the vortex when they re-achieve resonance as a result of the change in the dominant frequency of the vortex, in accordance with the evolution of the wave function. Non-resonant probability wave frequencies may leave the vortex altogether and form entirely new vortices, or join another vortex, or go back to the quantum energy jitters of the zero-point field. The vortices dance an elaborate tango to the flowing melody of their ever-changing frequency resonances as allowed by the uncertainty principle, whether they are in Irvine and Kleckner's water, Lord Kelvin's atom, Bose-Einstein cold condensates, tornadoes, galactic spirals, microtubules, societies, or in the human body.

Shwarzschild's solutions to general relativity theory equations also uncovered the prediction of wormholes. Wormholes are supposed bridges connecting two different regions of space-time. No wormhole has ever been found.

The gravity of the impact of Einstein's general relativity theory deepens with increasing formulations of solutions to its equations and inquiry into its predictions, all of them successful so far. A fresh approach might still fulfill the wormhole expectation. For all we know, the mechanism could be that of electron tunneling and entanglement of waves on a grand universal harmonic scale.

12
Dark Matter, Dark Energy

0.0000000000000000000000000000000000000000000000000000000000
00000000000000000000000000000000000000000000
00000000000000000000000000000000000001

Dark matter? Dark energy? They sound ominous! What are they? These questions I would have asked not too long ago, at the time when I realized that my friends and I had fallen sadly behind the march of accumulation of mind-bending and wowing knowledge in the last century. Although we have benefited from some of the fruits of technology, we have scarcely had a chance to enjoy intellectual stimulation that is offered by great minds of our time in a language we could easily understand.

The above outrageously small number was simultaneously discovered in 1998 by two independent teams of astronomers— Saul Perlmutter and his team at the Supernova Cosmology Project and Brian Schmidt of the international High-Z Supernova Search Team—who had been separately tackling the awfully tedious task of measuring the distances and redshifts of galaxies by utilizing light radiation from exploding stars of a specific kind, the type 1a supernova.

It is difficult to comprehend the above number with its many zeros. We can easily accept one-tenth, one-thousandth, or even one-millionth of one, but one-gazillionth? This ridiculously long string of zeros after the decimal point and before the number one, this miniscule difference from zero, consumed Einstein so. He thought it was a mistake, but it refused to vanish graciously.

When he formulated his elegant general theory of relativity, he realized that without introducing a fudge factor in the form of a "cosmological constant" to his equations, his theory would predict a universe that was not static, a universe that kept on expanding. This was in 1917, and Newton's stable, unchanging universe still held sway. If you stopped gazing at the stars one night, you knew that they would still be up there in their specified places at the appointed times for the next one thousand years. It was predictable, stable, and comforting.

Just as Goldilocks would have her porridge neither too hot nor too cold, Einstein's Newtonian universe had to be just right. He would not have a universe that could potentially allow an uncontrolled expansion. He had to make it so that the pull of gravity exerted by matter in the universe was countered by just so much expansion exerted by a repulsive antigravity force thus far unknown—a non-zero cosmological constant—in order to reach perfect equilibrium. Otherwise, either gravity would pull matter together in a tiny clump, or general relativity would drive the whole universe outward to a runaway rendezvous with oblivion.

Thus, ever so reluctantly, no matter how mathematically beautiful his equations were, and no matter his intuition of the theory's truth, the prediction of expansion had to go, and he factored into his equation the ever-so-tiny cosmological constant. He would, in just over a decade, call this decision the "greatest blunder" of his life.

A TINY NUMBER FOR AN EXPANSIVE ENERGY

In 1929, Hubble shocked the scientific world with the publication of his observations that the universe was expanding. The news capped a new era in cosmology—other sciences and technologies having marched on since Einstein captivated the world with his general theory of relativity.

Prior to Hubble's announcement, careful measurement of the redshifting of light from far-flung galaxies suggested that some galaxies were moving closer to the Milky Way, and others were moving farther away. Redshifting, which arises through stretching of wavelengths by time and distance, is based on a discovery by Christian Andreas Doppler (1803–53), an Austrian philosopher, mathematician, and physicist. He used the term "redshifting" to denote the change in light frequency toward the slower and redder spectrum, with the lengthening of the waves as the source moves away from the observer, as defined by the Lorentz transformations briefly mentioned in the previous chapter. The redder the shift, the faster the source is moving away. In cosmology, this situation could only mean that the galaxies in the Andromeda spiral nebula (now recognized as a galaxy), and other spiral nebulae that Hubble studied, were spreading away from their original positions and that the universe was actively expanding. (The radar speed gun used by highway patrol officers is a Doppler-based technology.)

With this background knowledge and armed with a one-hundred-inch (2.54-meter) reflector telescope, Hubble studied the distant galaxies. He meticulously measured the distances, wavelengths, and redshifts of the starlight that each galaxy emitted and painstakingly mapped them. What he discovered on the finished map was startling. The galaxies appeared to be flying away from one another. The farther apart they flew, the faster their speed of recession from one another, a rule of celestial behavior known as Hubble's law. (Remember Lemaître's scientific paper on cosmic expansion, published in French, that predated the introduction of Hubble's law?)

Cosmologists have estimated the universe to be expanding at a rate of 5 to 10 percent every one billion years. A calculation of the age of the universe according to this rate of expansion yields a number consistent with the ages of the oldest stars in the darkest night skies, further supporting Lemaître's big bang theory. Edwin Hubble's outstanding work sealed the fate of the cosmos and should have merited him a Nobel Prize. Unfortunately, only later, after his death, did the award committee consider giving the highly coveted prize to achievements in astronomy. The Nobel Prize is not given posthumously.

Einstein greeted the announcement of Hubble's discovery with excitement. It was a clear vindication of his original general theory of relativity before he "blundered" by including the cosmological constant. Because he had thought it was a mistake, he had eliminated it from all his field equations, and some physicists had also followed suit. They stopped using the tiny but non-zero constant in their equations, and for a while, it seemed that they could discard it with nary a whimper. But like a persistent case of tinnitus, the cosmological constant wouldn't completely go away. A few physicists found the tiny number to be actually necessary in some of their field equations.

Edwin Hubble's distance measurements, as those of other researchers before him, indicated the presence of a powerful something to drive the expansion of the universe, fueling the flight of components of cosmic matter away from one another: a certain repulsive and as of yet unseen energy. It was called "dark energy" for want of a better term and for the fact that it is invisible. It is without doubt, one of the most important puzzles begging a solution today.

Another earthshaking discovery came in 1998. The two teams who calculated the infinitesimally tiny number of Einstein's cosmological constant had been working independently of each other, and in the very same year, each team published the same conclusion from their separate studies. By analyzing the spectral lines from distant supernovae, they found that not only was the universe expanding, but also that its rate of expansion was actually accelerating, consistent with Lemaître's original hypothesis of an ever-increasing cosmic expansion. The two teams were awarded the 2011 Nobel Prize in Physics.

This discovery implies the presence of an unseen repulsive force that pushes galaxies away from each other against the pull of gravity, a very mysterious force uniformly filling all of space, exerting influence in a constant manner even as the pull of gravity gets ever fainter as galaxies move farther out and away from one another. This ever-present repulsion, steadily competing against weakening gravitational attraction, could account for the observed acceleration of cosmic expansion.

Lemaître had proposed in 1934 that Einstein's cosmological constant could be attributed to energy present in the vacuum of space. This constant could account for just the right amount of energy responsible for the expansion of the universe.

The cosmological constant and the dark energy fueling the accelerated expansion seem to be the same thing. This mysterious dark energy of not-quite-a-vacuum space amounts to reportedly 120 orders of magnitude more energy than that contained in the entire visible universe. The tiny numerical value beginning this chapter is now generally accepted to be Einstein's cosmological constant, his lamented "greatest blunder." One has to admire a great genius whose blunder reveals an even greater genius.

UNSEEN MATTER IN THE DARK VOID

Half century ago, dark matter caught the attention of cosmologists and astronomers and since then, it has been the subject of intense searching. Nobody has seen dark matter. No dark matter particle has ever been identified in theory or in experiments. Dark matter neither absorbs nor emits electromagnetic energy, the carrier of information on which experimental equipment and everything else in this experiential world depend. All support for the existence of dark matter has been indirectly gathered from its gravitational effects on its surrounding ordinary matter. The true invisibility of this material to the naked eye and to high-tech instruments is why it is given the name dark matter.

In March 2013, NASA reported new figures from the European Space Agency's Planck space mission on the composition of the mass of the universe, stating that 68.3 percent of the universe is dark energy, 26.8 percent is dark matter, and a mere 4.9 percent is normal matter. In January 2012, an international team of researchers working on the Canada-France-Hawaii Telescope mounted on Mauna Kea in Hawaii published the results of a five-year study of the gravitational lensing effect of some ten million galaxies—a number that boggles the mind, especially because when Hubble started peering into his one-hundred-inch telescope, only a dozen or so galaxies were visible to him. The results from Mauna Kea showed the largest map, thus far, of dark matter ever created: a composite picture of a web intertwining with the familiar visible matter generally considered to be the normal contents of the universe.

Cosmologists have postulated the halo of dark matter around galaxies to explain the puzzling spin of the beautiful galaxy Andromeda. The stars of Andromeda spin at a uniform and dizzying speed from the galactic center to the periphery. Their consistent velocity of rotation is incompatible with Andromeda's observable mass, the ordinary gravitational pull of which, reportedly by any form of calculation, dictates a much slower rotation of the stars in the periphery of its galactic disk. Ordinarily, the speed of rotation of a galaxy increases from the center to a maximum speed located in a region a few kiloparsecs[11] from the center, and then the speed is supposed to steadily fall the farther out it is measured.

---

[11] A kiloparsec is one thousand parsecs, one parsec being around 3.26 light-years or 30 trillion kilometers (19 trillion miles).

If there is a halo of dark matter, then the combination of the halo's gravitational effect and that of the visible matter in the vicinity of the galactic center could be enough to slow down the rotation near the center. This slowing effect could happen without affecting the speed of the peripheral stars, which would then appear to be spinning inordinately fast. Another proposed explanation is that dark energy is fueling the unusual speed of Andromeda's peripheral stars.

If an inconsistency has been trying to catch your attention while reading the above discussion, then your intuition is right. All this talk about dark energy and dark matter has treated them as if they are independent and discrete entities, even though, according to Einstein's special theory of relativity equation, $E = mc^2$, and Planck's $E = h\nu$, they are interconvertible. Mass is converted into energy or light frequency and energy into mass. $E = h\nu = mc^2$. They are one and the same thing, only expressed in different forms—in adherence to wave-particle duality. If this is true, then why should there be separate percentages for each?

ENERGY AND MATTER OF THE ZERO-POINT FIELD

While terminologies such as "backdrop ether" or "unseen fluid" filling all of space have gone by the wayside, the notion of an all-pervading unlimited energy has taken and dominated their place. At the extremely frigid zero point (noted in the previous chapter), space is expected to be empty, as all known particle energetic activity stops at this point. All known energy has been removed, and this space should be a vacuum. It is not.

Scientists have found the supposedly empty zero-point field to be full of irremovable energy jitters. The field is a sea of infinite energy teeming with undulations endowed with frequency, direction, polarization, and momentum. The postulate is that the zero-point field is fully suffused with electromagnetic fluctuation. Accordingly, this supposition means that the zero-point field is an electromagnetic field. The zero-point field is energy. The zero-point field is light. Morley and Michelson could not have found it, because they were looking for the luminiferous ether, for light, while also using light as a measuring device. This field of light frequencies, the zero-point field, is consistent with ancient accounts of such a state.

In the book *The Corpus Hermeticum*, Hermes Trismegistus, who was probably in a transcendent state, reports on how the Nous (mind) of the Supreme showed him the answer to his wish to know God and the nature of things. As Hermes wrote,

> all had become light, a gentle and joyous light; and I was filled with longing when I saw it. After a little while, there had come to be in one part a downward moving darkness, fearful and loathsome, which I experienced as a twisting and enfolding motion. I saw the nature of the darkness change into watery substance, which was indescribably shaken about.... There was sent forth from the watery substance a loud, inarticulate cry; the sound, as I thought, was of the light.

The energies in this zero-point dimension are of such high frequencies that they are outside of ordinary human apprehension and way beyond the capabilities of our machines. Scientists have postulated that in the realm of the very tiny, the quantum weirdness of the waves of probability allows the eternal presence of energy and its ceaseless, ever-so-transient crystallization into ever-so-tiny wisps of matter in the zero-point field: energy that is crystallized in a super-transitory moment and gone in the next.

Doesn't this imply that dark matter and dark energy are merely different manifestations of the same energy in the zero-point field, and that the confusing individual percentages science provides only reveal the dreadful inadequacy of current tools of measurement and the sad poverty of modern humans' understanding and linguistic expression? Quite possibly, and I am only presenting my humble thought: the sum, the totality of unmeasured zero-point energy vibrations, is what we are calling dark energy, and their ever-so-transient materializations that escape direct measurement by current scientific instruments are what we are calling dark matter, and the rest is the perceived matter: us and our world.

A different theory, based on the frequencies and resonances of the quantum "foam"—the zero-point energy field of intertwined frequencies—might someday provide the answers, but only theoretically and indirectly, as they are beyond the reach of our instruments.

$E = mc^2 = h\nu$. Interchangeable unseen dark energy, perceived matter, and frequency, each a manner of emanation of infinite light occupying all of space and unchanging through eternity—simply add up to a tantalizing embodiment of a mystery thickly veiled from ordinary human discernment. All we know is that the evident and awesome effect on our experienced world of sticks and stones vexes, teases, and thrills.

13
Consciousness

Consciousness can reduce even the most fastidious thinker to blabbering incoherence.
—Collin McGinn

Unus mundus: one world. Universe: one song. Consciousness: our singular apprehension of the universe. Consciousness does not exist in the plural. It is our whole and total recognition and understanding of our reality, internal and external; our feelings, aspirations, and sense of morality and duty; and everything that our five senses and extrasensory perceptions pose to us at any one moment. There is no separate consciousness for this and another one for that, although it may acquire different tones, such as political, social, or ecological; or varying degrees, such as dulled or heightened; or quality, such as pleasurable or disgusting. Consciousness is something that humans do not share with machines, artificial intelligence specifically.

Consciousness is "all or nothing at all," as Frank Sinatra sings to me from my smartphone. This is like Schrödinger's cloud of probability wave function, wherein the potentiality of the electron is anywhere and everywhere and yet collapses as a whole into a particle without leaving a single frequency behind. The ancient Hindu texts written in Sanskrit, the Upanishads, describe four states of consciousness: aware wakeful, dream, deep sleep, and transcendent meditative. All of them are amenable to scientific scrutiny by technologically advanced methods of functional brain imaging and electroencephalography.

Through all its states and all the assorted flavors that it acquires in each and every personal human experience in life, consciousness is absolute truth.

A singular consciousness is resistant to easy labels and eludes simple definition. Great world thinkers from the ancient days to the present have advanced strong positions, depending on their points of reference, be they physical or metaphysical. Quite a few major philosophies have considered the nature of consciousness. While none of them contradict the fact that consciousness exists, they do differ in their explanations of how it appeared on the scene in the history of the universe, or how it comes about in one's own particular world and how it affects the material body, the brain in particular.

Everyone acknowledges that one is conscious, since the very act of denying is itself done in consciousness. Common sense affirms to us our own conscious experience. For instance, I am aware that right now I am thinking hard about this subject matter. Also, common sense tells us that the mind affects the body. When I visualize my cousin's face when he is telling a joke, I chuckle merrily to myself. He is not actually present, being his funny self, but he exists in my mind as if he were right in front of me laughing, and my feelings and muscles respond to my mental image of him.

Many views on consciousness have been introduced over the years. One is panpsychism, a philosophy espoused by Plato, who believed in a world soul or anima mundi. Panpsychism is a belief that everything in the universe, from the largest to the smallest, has some form of consciousness. The philosophy has been incorporated in those Eastern religious teachings that maintain that everything in the universe, from human beings to trees and minerals, is inseparably connected to a dynamic force field called qi, *prana*, universal intelligence, source energy, Allah, God, or pure consciousness.

The Hermetic principle teaches that life manifestations come from "Source" consciousness or the mind of "the All." These manifestations are arbitrarily grouped into three indistinct planes as physical undifferentiated matter, the mental, and the spiritual. Subtly and seamlessly, they merge into one another, since they share the same intrinsic properties governed by the same laws from their common Source.

According to Hermetic teaching, a harmonic correspondence necessarily exists between all three planes of emanation from source energy, and that correspondence follows mathematical principles. Material manifestation gets more complex the higher up it goes within the same plane and all across each and every increasingly higher plane of emanation. Geometric crystallization gets progressively differentiated—ever richer and more elaborate—following the rules of harmonics. This makes sense because, as we understand from physics and chemistry, when two or more atoms join together by sharing electrons, a very different and more complicated harmonic composite results.

For example, two simple gases, oxygen and hydrogen, join together to form a far more complex liquid, water. This response to energy addition or deletion is repeated through a hierarchy of materializations, such as molecules, crystals, and human bodies. At each higher level, an increase in the complexity of their materiality occurs such as: their appearance, their particular reactions to energy stimulation, and their exclusively individual perception of their environment.

At this point in our accumulated knowledge and experiences, the quantum foam of zero-point energy may be correlated (though may not be fully equated) with the old Hermetic idea of Source and the Judeo-Christian creative energy of sound ("Speech" and "Word"). The old concept of fundamental energy agrees with the quantum-physical principle but may go deeper than that—so deep as to be beyond comprehension by the human mind.

Everything manifest starts with a focal increase in vortex energy within the quantum foam of fluctuating energy jitters. A particle such as an electron gains energy density and crystallizes ever so transiently in one tiny moment, only to be annihilated in the next by the positron. This arrangement would then be true with a composite of particles, from the simplest mineral to the most complex of beings, the human being. An increase in the density of energy in the cloud of potentialities can materialize as someone's image of a rock or a favorite face upon conscious human observation.

Each and every one of us is a harmonic vortex, a musical black hole of high energy concentration within one source singularity.

Human consciousness emerges here, when human energy amplifies the focal vortex energy of the observed entity and allows the amplified energy to break through the random jitters and materialize in the mind—ever in enduring oneness with Source consciousness. The idea that human consciousness starts from the probability densities of source frequency defines a clearer path than if one has to pick a specific point of insertion of consciousness anywhere in the course of the evolution of matter, from the simplest mineral to the most complex being. There does not seem to be an opportune juncture, beyond the zero-point field, for human consciousness to appear along the way.

We can say, following this line of thinking, that consciousness is present in the cloud of probability waves of the quark, but the quark's consciousness is simpler than the atom's, the atom's is simpler than the stone's, the stone's simpler than the plant's, and everything else's consciousness is simpler than the adult human's, which is the most complex, as far as is currently known.

The universe could also be conscious in a far more advanced way, although we may never know the truth through scientific experiments. The knowing of a conscious universe has reportedly always been and exclusively experienced in the transcendent state of bliss of saints, adepts, and mystics, and during rare unexpected moments of sharp clarity and all-embracing unity experienced by people in what otherwise would be ordinary life circumstances.

COMING TO TERMS WITH CONSCIOUSNESS

There are other schools of thought on consciousness as well. One popular view is the reductionist-materialist (in the context of material objects) or physicalist (in the context of natural physical laws) hypothesis, adopted by most in the scientific community. This school maintains that consciousness is solely the result of complex computational biochemical and molecular transactions in the brain that have developed during evolutionary adaptations. According to this view, consciousness is only a construct of some eighty-six billion cells of the human brain called neurons and is entirely of bodily origin, not a primary and intrinsic property of nature. In this view, consciousness arises out of the brain's activity without any need for non-material involvement in the evolution or function of consciousness; all of what we discern—our total reality—is fully and wholly birthed from biochemical processes in the material brain.

A problem arises from this materialist stance, though, because biological chemical processes in the brain cannot pinpoint the source of will and the freedom to choose, nor can they explain such non-sensory-dependent qualities as ethics, empathy, altruism, ambition, and the like. For example, right now, if the biochemical mechanism in my brain is geared toward writing this book and then is slightly interrupted by a mild hint of thirst, the biochemistry of my body does not all of a sudden switch from thinking about words to moving my skeletal muscles for me to walk to the refrigerator. No. First, I have to make a decision to get up and get a glass of water, and then the brain cells in charge of my muscles get fired up so I can move.

Neither can the materialist viewpoint explain my freedom to change my mind, nor my willful act of disregarding my thirst in favor of the stream of words in my thoughts. And of course, the materialist view cannot explain where and in what part of the numerous and complex processes in my brain involved in writing this book can I exert my choice to stay focused on writing instead of going for my glass of water.

There does not seem to be a special "stop and go" corner in the materialists' idea of consciousness at which at "stop," free will can insert itself, and then at "go" proceed to make something part of one's reality. These shortcomings of reductionist materialism and physicalism are unacceptable to many people. Mind-body dualism, which states that consciousness exists independently of the brain's chemical and electrical mechanisms, is often believed to be an innate property of the universe that somehow has a causal relation to brain function and, therefore, to one's awareness of material existence. Such dualism is in the true nature of an omnipresent, omnipotent, and omniscient God embraced by the spiritual community and is beyond the reach of scientific inquiry.

Since consciousness in this concept is an entity different from the anatomy and the biochemical function of the brain, the idea provides a point in space-time for the introduction of will and decision-making. The problem with this view over the years has been that scientists have not been able to offer a mechanism—the "how"—by which consciousness, independent of and distinct from the brain anatomy and chemistry, could instigate the biochemical computational activities in the brain that construct someone's experience of reality. Recent scientific advances have revealed such a mechanism, as we shall see shortly.

Somewhere in between dualism and reductionism philosophies is the idea of consciousness as put forward by Alfred North Whitehead (1861–1947), an English mathematician and thinker. This concept agrees with dualists' notion about the existence of consciousness as separate and distinct from the anatomy and biology of the brain but offers that consciousness is also a function of the experience of momentary spatio-temporal physical biological events in the brain tissue: a notion parallel to that held by the French scientist and thinker René Descartes (1596–1650).

Descartes believed in a separate existence of the material physical aspects of nature—the corporeal substance he called res *extensa* (Latin for "extended thing")—from that part of nature comprised of thoughts, feelings, and ideas that cannot be described in mathematical physical terms: his *res cogitans*. He thought that both aspects could interact with each other inside a human brain.

According to Whitehead, mental "events" or "occasions of experience" are like a succession of the individual still frames of a movie film in a continual stream of creation and dissolution. One can relate this phenomenon to what happens in the quantum foam of the zero-point field in modern physics. The phenomenon constitutes the nature of the world experienced by each and every form of matter in the universe, similar to Plato's panpsychism (he called such forms "individuals"), from the simplest particle to the most complex and highest form of all compound matter. As in panpsychism, the simplest entity in the universe has the same occasions of experience, thus indicating an individual creative aspect of the perception of reality.

Members of *Homo sapiens*, from the vantage point of our paramount complexity compared to the simpler forms of matter, most acutely experience the richness and subtlety of our reality. And our ability to exercise our free will and our capability for self-determination are unsurpassed in nature. Whitehead's type of panpsychism is differentiated from the rest of the genre by the term "panexperientialism." The reason is that in this philosophy, everything must have an inherent "mysterious reality in the background, intrinsically unknowable," in Whitehead's words. This reality is outside the material state and is undeniably spiritual.

Panexperientialism has a quantum biological counterpart in the orchestrated objective reduction (or Orch OR) hypothesis—the resonantly directed collapse of the wave function in the brain—proposed by Sir Roger Penrose (b. 1931), and Stuart Hameroff (b. 1947). Penrose is an English physicist, mathematician, and philosopher, and Hameroff is a professor emeritus of psychology and anesthesiology at the University of Arizona in Tucson, where he is director of the Center for Consciousness Studies. In their view, orchestrated objective reduction results in a "moment of proto-consciousness," a purely physical phenomenon that takes place according to the laws of physics before materialization and is thought to be more primitive than the state of consciousness that creates the observer effect. In this sense, proto-consciousness is a reality operating behind human conscious awareness.

In a 2013 interview with Deepak Chopra, the Indian-born American physician, author, public speaker, and Ayurvedic medicine practitioner, Hameroff likened proto-consciousness to an ocean, the surface of which is riddled with individual tiny waves and ripples that have not quite broken. The emergence of consciousness is like the crashing of fully formed waves on the shore, or the breaking of foamy whitecaps in the deeper sea, equivalent to objective reduction or the collapse of wave function. In this ocean is the oneness of all wave functions, all vibrations, and consciousness: no different from the oneness of the water that ripples or breaks at the shore with the whole of the sea. In the same way, proto-consciousness is the "pure consciousness" concept of Eastern religious tradition: what the majority of us call Spirit or God.

Immanuel Kant (1724–1804), a German philosopher and one of the most influential thinkers in Western philosophy, upheld the same idea. He made a distinction between "phenomena," a term that is inclusive of a human being's total experience of reality, and "noumena," the things-in-themselves reality beyond the perceived world responsible for the phenomena.

QUANTUM RESONATING WITH CONSCIOUSNESS

Times are a-changing. Discoveries like quantum entanglement, the cloud of probability waves, wave-particle duality, frequency-orchestrated Bose-Einstein gaseous condensates, and quantum communication in plants and animals make plausible the existence of consciousness that is independent of the anatomy and biochemistry of matter. Recent research findings on quantum processes in the brain have shown how consciousness outside the brain anatomy brings about brain activities to faithfully effect our perception of the world around us.

A 2014 study by Subrata Ghosh and colleagues demonstrated the presence of resonant frequencies in the microtubules of brain cells that orchestrate the cells' oscillations. The team found that these vibrations instantaneously transfer information resonantly and wirelessly across the whole brain in a process akin to the binary computation of one and zero bits that occurs in digital computers. In effect, they found that the microtubules appeared to be actual live quantum biological computers.

Resonant frequency microtubular activation brings quantum entanglement into the picture of brain function and human consciousness. Our flow of conscious experience starts when different frequencies are input into our sensory and extrasensory mechanisms from our internal and external environments as simple yes or no questions of frequency resonance, harmonically entangling them with our own vibrations. They are then transmitted throughout the nervous system, including the brain, where they are incorporated and experienced: all instantaneously happening in accordance with the principles of frequency, resonance, and harmonics.

According to the scientists mentioned above, a mechanism exists for the useful resonant sequencing of information in the form of wide-range resonance frequency bands arranged in fractals in the whole brain. (As discussed earlier, a fractal refers to the unceasing repetition of geometric or numerical patterns in any scale.) Each band acts like a built-in internal timekeeper, since frequency is typically measured in cycles per second.

From what we know about interference, frequencies exist within beat frequencies, which in turn exist within lower beat frequencies, and so on and on, in the manner of the nesting of spheres and tetrahedrons, of dolls in Japan, matryoshka dolls in Russia and boxes in China, and of information nesting in computer science. Arranged resonantly, the frequencies never get jumbled up and information makes sense, whether inside a single cell or in the entire universe.

The accomplishment of optimum sequencing of resonant frequency information depends on the highly complex network of microtubules inside the cells of an organism, especially brain cells. These microtubules appear to be explicitly structured for optimum quantum reception and transmission of frequency signals between parts of the cell and between cells, organs, the whole human organism in all its complexity, and all of nature.

Researchers have observed microtubules to automatically self-assemble into a microtubular vortex and to circulate clockwise and counterclockwise. Inevitably, the smooth collective motions of individual microtubules result in the formation of structured lattices. This spontaneous formation is commonly observed in nature as well—as seen in the seemingly orchestrated motion of a school of fish or a flock of birds—and had not been sufficiently explained before these recent discoveries.

Electromagnetic communication between the structured water inside the microtubules of individual group members and the surrounding environment allows the members to communicate with one another and with the group as a whole instantaneously. Frequencies in the surrounding environment and in microtubules of individual members of a group are resonantly entangled as a whole and ordered as fractals of band frequencies, allowing instantaneous sharing of information, thereby promoting the orchestration of harmonic group formations and synchronization of group collective movement seen in the animal kingdom.

Harmonically integrated fractal patterns found in the small tubulins and their structured water content in megahertz and gigahertz ranges are self-similarly found in the microtubules that they comprise, in the kilohertz and megahertz ranges. The fractal patterns are repeated in the bigger microtubules—in increasingly lower frequency beat ranges.

The repetition of microtubular fractal frequency pattern extends to neurons and to axons, the long microtubular tails of neurons that connect with other brain cells through the synapses. The self-similar pattern continues on to clusters of brain cells and to the whole brain, which, anatomically, is itself a fractal. This fractal replication facilitates the simultaneous collective firing activity of neuronal cells, axons, and synapses in different areas of the brain, including different lobes, even though the lobes are anatomically separated by deep grooves.

In music, a beat occurs when two waves meet and experience different degrees of constructive and destructive interference at certain intervals. The precise patterns created by the constructive and destructive changes in their amplitudes produce another wave of a lower frequency, a new and slower oscillation: the beat. The fact that individual frequencies exist as fractals makes it natural for them to be electromagnetically transmitted via beat frequencies globally and instantaneously.

Scientists have found evidence that the EEG represents the brain's beat frequencies. A single neuron could potentially be connected to ten thousand other neurons by some quadrillion synapses: a one-trillion-bits-per-second processor if it were a computer. The synchronous firing of neurons had been observed in the laboratory before, but the mechanism for their simultaneous activation remained unknown until recently.

A MEETING OF MINDS

The growth of an organism, such as a fetus, requires the multiplication of cells, termed mitosis, within the mitotic spindle inside the cell nucleus. To achieve this, the forty-six chromosomes of the human being replicate to form two identical copies each: two twin DNA strands. Initially, they are all tangled up in a mesh of DNA and protein, requiring fine disentangling before they can be accurately separated and the cell exactly divided into two completely identical cells.

The process involved in mitosis is the epitome of exquisite quantum biology in action. All of its precise steps are orchestrated by resonant vibrations of the crystalline water component of the microtubular "water-tubulin-MAP-kinetochore complex," which behaves as one molecule. Otherwise, chromosomal deletion, duplication, translocation, and other rare forms of genetic defects occur. The familiar Down syndrome (also known as trisomy 21), a common form of genetic material translocation, results from the cell having three complete or partial copies of chromosome 21 instead of two: a failure of the critical procedure of chromosomal genetic division.

It was while Stuart Hameroff was a young medical student in Philadelphia in 1970 that he spent some time in cancer research, particularly on the multiplication of cancer cells. He noticed that the mitotic spindle, a collection of microtubules inside a dividing cell, appeared to display some sort of innate intelligence—in other words, some form of consciousness.

Around this time in physics, Roger Penrose, whose research at this point was focused on the nature of consciousness, was developing and proposing the theory of a non-mechanistic, non-computer-algorithmic-like consciousness that goes beyond what computers and artificial intelligence can do. He was convinced that quantum mechanics could explain the process, and he put his thoughts forward in his 1989 book titled *The Emperor's New Mind*.

After reading the book, Hameroff realized that Penrose had the quantum-physical equivalent of his biological research discoveries on consciousness and brain microtubules, and so he contacted Penrose. Their meeting began a long-running collaboration to develop a quantum theory positing that human consciousness evolves through orchestrated objective reduction, known as the Orch OR theory. The process involves a harmonically directed "reduction of the wave packet," or the resonantly guided collapse of Schrödinger's wave function at the microtubular level into macroscopic electro-chemical processes in the brain cells and their synapses. This results in what the two researchers call "moments of conscious awareness," after which conscious experience may or may not arise. Schrödinger had originally thought that the wave function could essentially deny reality. And Einstein had lamented the uncertainty of the wave function: "I like to think the moon is there even if I am not looking at it."

Einstein didn't have to worry. The moon is there and it is not. An object as a wave function is real in the non-manifest physical realm of vibrations. At the same time, the object is an unreal illusion of material existence in the mind. We are a vortex of vibration—a soul—having an awesome bodily experience.

The Orch OR hypothesis is supported by the work of other scientists. In a 2014 paper published in the journal *Physics of Life Reviews*, Ghosh and colleagues wrote: "Hameroff and Penrose have rightly argued . . . that wireless communication of axons via resonant vibrations around a hundred micrometers diameter domain alleviates the biggest criticism of the Orch OR proposal. The orchestration of resonant vibrations occurs globally between all neurons across the entire brain." From what we have discussed, it is evident that this wireless synchronization and instantaneous transmission of resonant information all across the brain is more fundamental than the initiation of electrical current and its transmission through changes in brain cell electrochemical levels. Resonant information transfer underlies all brain activity.

Beyond these scientific findings, though, Penrose and Hameroff maintain that "something else" operates underneath this level, something not quite clearly defined. Penrose has suggested that this "something else" works in a manner similar to a background quantum computation that proceeds in the conscious mind in accordance with Schrödinger's wave function equation. The moment a certain threshold is reached, the wave function collapses, and the quantum computation stops.

The threshold is reached when frequency resonance is achieved between the probability waves of both source energy and the conscious observer. A new harmonically entangled pattern is established and crystallized in the observer's mind. This event results in a distinct and intimately personal human conscious experience of reality that is connected to the aforementioned "something," the proto-consciousness of Penrose and Hameroff, which the latter likens to the concept of the unfathomable pure consciousness found in Buddhism.

This underlying "something" is what we commonly refer to as pure mind, pure consciousness, universal intelligence, source energy, and God, the immanent, boundless, and unknowable source of everything. To quote Chopra, "From it everything comes and to it everything returns."

According to the ancient Sanskrit text the Bhagavad Gita, the mind sees all things united with the ultimate reality, one's own self included: "Know this, O Prince: Of things created All are come forth from the seeming union of Field and Knower, Prakriti with Brahman. Who sees the separate Lives of all creatures United in Brahman Brought forth from Brahman, Himself finds Brahman."

Field and knower. Wave function and conscious observer. From their union comes forth the human mind's understanding of all things created, each and every single one in its own uniquely individual existence but united with and by Brahman: pure consciousness, or God.

DOES CONSCIOUSNESS CREATE REALITY?

Wisdom handed down through the ages has always proclaimed that reality, as we experience it, is mental and that the material world is brought forth from pure mind by our conscious thought. A Hermetic principle states: "the universe, as a whole, and in its parts or units has its existence in the Mind of THE ALL, in which Mind we live and move and have our being." Since we are one with eternal pure mind, the whole corruptible material universe and our mortal bodies exist only in our consciousness.

In this reality, the fabric of space-time is construed solely by the mind, as is true in both the dream and schizophrenic states. Space and time are not absolutely real and do not have distinct independent four-dimensional structural existence. This means that a red and pretty object swaying in the breeze and smelling like a rose is only a brainchild of one's imagination. There is no object called a red rose out there. But everywhere and boundless is the vortex of energy of a fragrant red rose eternally entangled with the frequencies of the conscious observer and source energy.

Quantum physicists proclaim this lack of absolute materiality in many ways. Heisenberg believed that the unreality of electrons or atoms—which exist only as potentialities or waves of probability until their contact with an observer—extends to large objects as well, such as that rose, that ball, that dog, that man, that moon, and this whole universe. As noted earlier, Pascual Jordan claimed that the act of observation not only disturbs what is to be measured, but that observation itself produces an effect. Schrödinger himself said that "every man's world picture is and always remains a construct of his mind, and cannot be proved to have any other existence." There is no real material object existing outside the observer's very own consciousness.

In their 2009 book *Biocentrism*, Robert Lanza, a pioneer in stem cell therapy and cloning, and Bob Berman, an astronomer and science editor, hypothesized that consciousness creates the universe, and that the mind's conscious perception makes an individual's reality. The mind is pre-spatio-temporal and exists independent of and before the space-time constructs of brain processes that order information so that it can be experienced as reality. Everything exists in the mind alone, according to the authors: "The universe bursts into existence from life, not the other way around." The universe is exactly the way it is, perfect for supporting life, because life itself made it that way.

But the moon is really, truly there, whether or not there is a human being or any other animal (or maybe even plant or mineral consciousness) to collapse its wave function. It is there, in the zero-point energy field, as clouds of probability vibrations in harmonic arrangement according to Leibniz's mathematical and geometric relationship. Fido does not see his master's moon in the full range and hues of yellow, since his eyes' cone receptors are limited to blue-violet and a narrow range of yellow. On the other hand, he can hear frequencies that his master cannot. So, does he hear a certain song from the moon, and that is why he howls at it some nights? Does he have a specific howl that describes it in the same manner by which his master says "beautiful moon"? Nobody knows—as of yet.

The moon looks the way it does to the human eye because the wave function of the massive number of frequencies that make up its cloud of probabilities collapses upon encounter with an observer. This happens in your own lovely pair of eyes, the retinal rods and cones of which quantum-mechanically transmit the resonant frequencies of moonlight to your brain microtubules and are perceived by your mind as a beautiful moon.

The moon's information also undergoes many more occasions of entanglement (including free association with memories connected to that moon) during information transfer, processing, and integration into consciousness. Thus, my mental construct of that moon is really not exactly the same as that of my friend's.

Additionally, the sounds, words, feelings, and actions that ensue because of that moon create many more occasions of entanglement and wave function collapse into material earthly reality.

If the moon induces a lover to sing to his sweetheart with all the emotion he can bear, then the vibrations from his intention, his voice, his heart, his whole body, and the very lyrics of the song affect the air and his surroundings, both immediate and remote, including his beloved's eardrums, nervous system, whole body, and consciousness. The effect of the song on his darling, on the still soft night air, and on the birds listening in silence to his sweet serenade, are physically and vibrationally real. All of them are participating in a quantum-mechanically tangled embrace with him and the moon. The moon's vibrations, thereby, are hopelessly interlaced with everything in his reality, including his darling, who doesn't even see it on account of being blind.

Consciousness orchestrates it all, instantaneously. According to the physicist Sir James Jeans, the "mind no longer appears to be an accidental intruder into the realm of matter ... we ought rather hail it as the creator and governor of the realm of matter."

Once again, we are reminded that the hauntingly beautiful sunset, the joyful sound of a bird's trill, the soft touch of a baby's fingertips, and the delicious aroma of fresh bread baking are not out there in objective existence. They are constructs of the mind, appearing as real as they do in all their vibrant and breathtakingly kaleidoscopic abundance.

14
The Holographic Universe and Princess Leia

Just as every portion of a hologram contains the image of the whole, every portion of the universe enfolds the whole. This means that if we knew how to access it we could find the Andromeda galaxy in the thumbnail of your left hand We could also find Cleopatra meeting Caesar for the first time.... Every cell in our body enfolds the entire cosmos. So does every leaf, every raindrop, every dust mote.

—Michael Talbot

Dennis Gabor (1900–79) was a British-Hungarian physicist and electrical engineer who worked at the British Thomson-Houston Company, and later at the Imperial College London as professor of applied physics. In 1947, while he was still at his first employment, his interest was centered on improving the electron microscope, then still in its infancy. Using a type of calculus invented by the French mathematician and physicist Jean-Baptiste Joseph Fourier (1768–1830) some 125 years earlier, Gabor applied mathematical equations, known as Fourier transforms, to convert images into waveforms. Then he exploited the same equations to convert the waveforms back to their original images. He stored the interference patterns of the waveforms in a special film and re-created the original image as a hologram by shining light through the film. He received the 1971 Nobel Prize in Physics for this invention.

The conversion of images back and forth is utilized in the current technique of sending signals to TV receivers at home.

213

First, the images are converted to different light-wave frequencies and then imposed on carrier waves to enable them to travel the airwaves, and into pulses of one and zero bits to travel digitally. If the digital transmission is electrical, then the ones are represented by high voltage and the zeros by low voltage. If the transmission is optical, then ones are light and zeros are dark. Upon hitting the TV receiver, the specific frequencies are then translated back to images, which are displayed on the television screen as the specific program being played (with accompanying sound).

The word "hologram" comes from the Greek word *holos* for whole and *gramma* for message. Holography is the process of encoding a special photographic film with fields of light frequencies—that is, information—as opposed to the capturing of lens images in ordinary photography. Without cumbersome equations, the following simplified description, using the *Star Wars* hologram of Princess Leia as an example, is how holographic images are created.

A beam of laser light is split in two. One of the pair is focused on Carrie Fisher, who plays Princess Leia. The other is beamed toward a mirror, which reflects the beam to another mirror, which then sends it to a spot in front of (not on) a very high-resolution photographic film: the exact spot where the beam reflected directly from Princess Leia also hits. This collision of beams results in images of concentric waves (Huygens's waves) that form an interference pattern (Young's pattern) that hits in a spread-out manner all over the photographic plate.

When a laser beam, or any strong light beam, is shone on the finished film, a perfect three-dimensional image of Princess Leia begging for the help of Obi-Wan Kenobi appears—not just the part of her that is directly hit by the light beam, but all around the whole of her, as if she were really there. This is one very interesting feature of the hologram, its three-dimensionality: a reflection of the spherical nature of Huygens's concentric waves.

We can actually examine the image from all angles—front, back, up, down, and sideways—as if casing out the actual person of Carrie Fisher. And then, when we reach out to touch her, she is not there. If we cut up the film in small pieces and illuminated the tiniest piece, the full image would still be visible in its bona fide form with not one detail missing, although the sharpness would be compromised the farther out the cut piece was from the center of the film. All information contained in the frequencies of the waveforms is preserved in every spot on the holographic film. Just as the tiniest part is contained in the whole, the whole is contained in the tiniest part as well.

The same holographic phenomenon applies in a fractal image. As discussed earlier, a fractal is a never-ending replication of a pattern in ever-smaller or ever-larger scales, achieving complexity and beauty through the simple process of self-repetition. A fractal's parts can be looked at under ever-greater magnification to show a miniature copy of the whole, with no end in sight: only the same self-similar picture repeated to infinity.

Overt examples of fractals abound in nature in the branching of trees; in the pattern of river tributaries, hurricanes, tornadoes, and typhoons as viewed from satellites; and in the forms of spiral galaxies, seashells, young ferns, flowers, cacti, and more. Subtle exhibits are also present in snowflakes and sound-induced water crystallization. In industry, fractal antennas are used for mobile devices such as wireless internet devices, cell phone routers, television, and GPS, Bluetooth, and other radio communications in order to achieve a wider frequency coverage in a much smaller and lighter mobile unit.

In the late 1960s, Karl Pribram (1919–2015) collaborated with David Joseph Bohm (1917–92) in the development of the holonomic theory of brain processing. Pribram was a prominent neurosurgeon and distinguished research professor of neuroscience at Georgetown University and Bohm was a physics professor at

Princeton University. Bohm closely collaborated with Einstein at the nearby institute and is widely recognized as among the most significant theoretical physicists of the twentieth century.

The holonomic brain theory is a holistic and analytic model of human brain processing as it relates to consciousness, perception, imaging, and attention, a theory that likens consciousness to a hologram. In 1969, Pribram hypothesized that information in consciousness is received, processed, and then stored by the brain in the holographic form of oscillations in the brain's web of fine dendritic (treelike) connections between brain neurons, contrary to the prevailing belief at that time that memory is stored in one anatomic location in the brain.

Pribram's model of holographic brain processing involves the large structures of the brain cells and their fine branches (dendrites), which form webs with one another; through these webs, the dendrites influence neighboring cells without requiring any actual contact. This processing happens as well in the dynamic electromagnetic fields surrounding the dendrites.

As a hologram, every bit of information is present in every piece of the pattern of wave interference and is available to every part of the brain, not to just one specialized area for memory. This picture of a holographic memory offers an explanation for the ability of a person to associate a piece of information with a certain stored memory, and to retain intelligence or recover memory after massive brain damage or surgical removal of substantially large portions of the brain.

Much in the same way that the light-interference patterns of the hologram take on cinematic life when played on a machine, consciousness takes on life in the conscious brain's fine structural oscillations, replete with varieties of colors, sounds, textures, smells, flavors, shapes, emotional content, duration, distance, and movement in each and every personal experience. This way of thinking removes the veil of mystery shrouding wave-particle duality and the enigmatic existence of all things in the universe, great and small. In the human being, the conscious brain interprets the frequencies that the brain resonantly apprehends. The frequencies are waves before orchestrated reduction in the brain occurs, and then particles or objects after—but existing only in the conscious mind.

Bohm believed that the brain works at the quantum level and that thought or mind as a quantum entity is entwined with the rest of the universe. His theory of "implicate" order (from a Latin root meaning to enfold or to fold inward) expresses the "unbroken wholeness of the totality of existence as an undivided flowing movement without borders," a movement as a whole as required by quantum entanglement. He used the Mandelbrot fractal set to illustrate his view of implicate and explicate orders of reality.

The implicate order enfolds or gathers information from the very tiny to the very large and encodes it in the complex patterns formed by the multiple wave interference of different interacting electromagnetic fields. The holographic whole gathers information and stores it as light-interference patterns forever. In this manner, information is never lost.

From the implicate order, things are unfolded unto the earthly space-time reality of the explicate order, each thing in its own particular region of space and time, separate from the regions in which other things belong. The endless participatory process of enfoldment and unfoldment as a whole, extending way beyond human comprehension and apprehension into infinity, is called "holomovement." The implicate order has all the information— the frequencies and geometric forms of every type of matter, from quarks to the whole known and unknown cosmos—ready to be rolled out at unfoldment and to be crystallized in space and time and become our perceived and treasured reality; just as the full image of Princess Leia was only waiting for a beam of light to give it life-like cinematic existence.

This idea of a manifest part contained in the whole and the whole contained in the part has been around for a very long time. Hermetic teaching espouses that the universe as a whole and in its parts exists in the mind of "the All." Judeo-Christian belief, as written in Acts 17:28 (KJV), maintains that "for in him we live, and move, and have our being; as certain also of your own poets have said, For we are also his offspring."

As such, everything that is—seen and unseen, heard and unheard, tangible and intangible—and everything that emerges from the enfolded whole through unfoldment is literally the offspring, the child of the Father, or the All, or the holographic implicate order. The offspring manifests as material life in human consciousness and is one with its Father as the Father is one with it, a notion that is truly compatible with and strongly supports the core of all Eastern and Western religious beliefs—if one studies them with open heart and mind.

It is this steadfast, intertwined, whole characteristic of fractal holographic existence that completely zaps away the preposterous spookiness and weirdness of quantum entanglement, the faster-than-light "action at a distance" that Einstein never accepted as happening without a hidden cause. Tweaking one tiny part of the holographic pattern immediately and expressly affects all other parts faster than the human-measured speed of light. This notion gives life to an old saying: "Pluck a flower in the garden, disturb a star in the heavens."

Although John Wheeler (Einstein's collaborator discussed earlier) said that nothing exists until it is observed, this statement may mislead if we do not carefully consider it. This idea applies only to perceived matter and disregards the fact that the frequencies that make up the quantum yes-or-no questions posed to the microtubules exist beforehand in waves of probability, before the act of conscious observation occurs. The energy frequencies of the zero-point field are infinite and eternal. Their unified wave function is harmonically entwined with the wave functions of the unique vortices of individual realities in the material world—all the objects and creatures of the universe—through eternity.

The waves of probability could possibly be perceived by a dog or a plant or even a stone in different ways, according to the manner in which the frequencies of the one wave function and its eddies of individual wave functions join and establish resonance with the vibrations of each and every perceiver—the dog's vibrations being different from the plant's, the stone's being different from either one, and all of them different from those of a human being.

Consciousness is often equated to spirit, the immanent, timeless, all-pervading, unchanging, infinite and thus unknowable energy. Consciousness is given inadequate expression in such monikers as quantum foam, energy fluctuations, zero-point field energy, and energy jitters. It is Hermes's mind of the All, Parmenides's one being, Hindu's Brahman, Eastern philosophy's pure consciousness and light, Judaism's the Being (whose name, YHWH, was considered too sacred to be uttered), Islam's Allah, Christianity's Almighty God, Native Americans' Great Spirit, the essence of *unus mundus*, Plato's world soul, Kant's noumena, Bohm's quantum potential, de Broglie's pilot waves, EPR's ultimate hidden variable, Bohm and Pribram's holographic whole, Whitehead's background mysterious and intrinsically unknowable reality, Hameroff and Penrose's proto-consciousness, and Lanza and Berman's pre-spatio-temporal mind.

The multiple references to this essence boil down to just one thing: frequency, energy, light. These terms, in turn, are only inadequate linguistic attempts at communicating, as best as people can to one another, this basic stuff orchestrating reality that passes human intellectual rumination and comprehension and lies beyond the reach of equations, illustrations, and words.

In some special interdependent way, consciousness informs matter and is likewise informed by matter. The understanding of fractals and the proposal that information exists as a hologram—wherein frequencies of the past, present, and future all coexist forever and are available for the human organism to access anytime and anywhere—supports the ancient belief in the reality of one pervading consciousness.

But that does not necessarily guarantee someone an everyday state of having supreme knowledge or extraordinary wisdom or of possessing other prodigious abilities. That does not seem to be a natural endowment in the ordinary life domain, at least for the bulk of us humans, probably because modern people have forgotten how to connect with our innate ability, or we have lost most of it through the long course of our evolution, or we are too busy to pay attention. With intention and effort, we may realize the depth of our loss enough so that we will develop an intense longing for this obscured property and make a firm commitment to recover it.

In any case, the premise of one consciousness does provide a plausible explanation for the astounding talents and skills of child prodigies and savants, which go way beyond what they could have quickly learned in their tender years. A familiar example of a savant is Kim Peek (the "Rain Man" of the movie of the same title), and that of a child prodigy is Wolfgang Amadeus Mozart, who wrote his first musical piece at five years of age and his first symphony at the tender age of eight.

The idea of one consciousness makes it easier to accept the almost-accurate knowledge and descriptions of events exhibited by remote readers and seers, abilities that go beyond their obvious limitations of space and time. The notion offers understanding of the depth and range of knowledge, wisdom, and inscrutable capabilities of those in a state of altered consciousness during transcendental meditation and spiritual bliss.

The concept of one consciousness makes sensible a person's intimately personal intuitions of deep knowing, rare and sudden startlingly crystal-clear insight, and precious "aha" moments, thus rendering them less subject to ridicule and humiliating dismissal as meaningless products of chance or wild imagination, or just plain trash and stupid superstition.

The existence of one consciousness easily allows for someone's near-death experience of observing events unfold around his body as if he were watching the scene as a separate entity from the lifeless heap of matter containing his physical brain. It may even explain the visible changes in the hands and feet of the stigmatic, whose vibration, in fervent prayer and bliss, resonates with the frequency of the crucified Jesus, which is eternally present in the cloud of probability waves of the zero-point energy field.

In an essay titled "Hospitality and Pain," which he presented at the Chicago McCormick Theological Seminary in 1987, the philosopher, educator, and Roman Catholic priest Monsignor Ivan Illich said that "compassion with Christ, for these late medieval mystics, is faith so strong and so deeply incarnate that it leads to the individual embodiment of the contemplated pain. The stigmata of Saint Francis of Assisi witness to the enfleshment of his faith in the Incarnate God who faces him from the cross."

Monsignor Illich called this enfleshment—the actualization in flesh and blood of the faith of Saint Francis as he gazed entranced, filled with full compassion on an image of the crucified Christ. If we analyze this from the standpoint of energy vibrations and frequency, we'll find that the saint's empathy with Jesus Christ was so strong that his vibrations resonated and merged with the frequency harmonics of the crucified Christ and His bloody wounds. In full compliance with quantum physical law, the multitude of vortices of their entangled frequencies became materialized and embodied as stigmata of the wounds of Christ on the saint's feet and wrists.

The phenomenon is in complete accord with Bohm's theory of an implicate and explicate order of fractal holographic reality: the holomovement of incorporation of frequency information from the crucified Jesus and from Saint Francis into the holographic whole and then, unfoldment of information into visible matter. The manifestation is accomplished through Hameroff and Penrose's harmonically orchestrated collapse of the waves of probability wave function upon reaching critical energy density and resonance in the microtubules of the brains of Saint Francis and other stigmatics and their conscious observers.

The above-mentioned non-rational cognitive phenomena abound in our real world of sticks and stones, albeit disregarded by science and its priests and converts. Of course, we can expect that most members of the scientific community will toss out any and all effort to try to connect extraordinary and mystical human experiences with acquired scientific knowledge as plain rubbish or pseudoscience (whatever that term may mean). Labels cannot and do not really nullify the truth. They do not matter.

What does matter is that many ordinary people, as well as philosophers and scientists, are truly seeking relevance between the extensive knowledge gathered within ivory towers and behind the walls of scientific academia and the reality they encounter. They are compelled by a nagging necessity for the dots to be connected, because they sincerely feel that the connection is there, and the distressing separation between science and their acquaintance with life just cannot remain forever.

The very real life experiences are there. The scientific facts are there. Marry the two with open heart and mind, and one just might appreciate life in a totally holistic way. This way of thinking makes sense to the logical mind. It feels right to the intuitive gut.

The concept of one consciousness, from which every frequency in the wave function emanates to resonate with the human mind into manifest reality of the moment, embraces the long-held oneness philosophy of the ages and the mysterious phenomena of quantum entanglement and observer effect. Maybe not in the same manner that this oneness was understood before, but in its fresh and clear discernment as a full and beautiful symphony. Each and every one of us and the whole universe is a vortex of harmonic vibrations, each a special eddy of melody in the music of one all-encompassing energy.

Resonating with the tune of pure consciousness, the human body ceases to be a mere mechanical compilation of atoms and becomes a conscious being. We are pure spirit experiencing a material body of our imaginings. According to the physicist Brian Cox, "We are the cosmos made conscious and life is the means by which the universe understands itself."

15
The Nature of Reality

As I was sitting in my chair,
I knew the bottom wasn't there,
Nor legs nor back, but I just sat,
Ignoring little things like that.
—*Hugh Mearns*

The current assessment of the atom is that it is mostly empty space and that the stone someone's toe just stubbed against is a composite of atoms of mostly empty space. And yet the toe, also a composite of atoms, did not go smoothly sailing through the stone. It got painfully stopped and bruised by it. The reality of the single atom and the reality of atoms comprising an object like the stone are two very different things to the human mind.

But if one thoughtfully considers the frequency dissonance between the electrons occupying the rigid valence shells of frequency of the atoms composing the stone and those of the human body, the idea becomes easier to understand. The toe had to be rejected by the inherent composite frequency of the stone's valence shell electrons and vice versa, which, being non-resonant, caused the toe to bounce off and hurt. That's the same way dissonant sounds hurt the ears, and it is the same mechanism by which a massive dying star explodes into a supernova from the strong repulsion by the electrons of its iron core.

In the song "High Hopes," performed by Frank Sinatra, a tiny ant with high apple-pie-in-the-sky hopes succeeds in moving a rubber tree plant. In the common way of looking at things, there is not a chance in the world that a mere ant could ever budge even a small seedling, much less an adult rubber tree plant. But who would know for certain? Maybe the little old ant, in its own way of perceiving its reality, did actually move that rubber tree plant some way, somehow. The ant's truth just cannot be seen from a person's ordinary standpoint.

Reality is an idea that stubbornly eludes satisfactory definition. It innately possesses a subjective quality that makes the experience of a set of information vary from that of an ant to that of a person's. Reality is rooted in the uncertainties of quantum probability waves and their vibratory connection with the observer who is having the experience, enmeshed with all the complexity of his apparatus of perception, his ability to process and integrate information, and his personal history of emotional, mental, and physical memories related to the information he has received. Objective reality does not exist in an individual experiential world. Surely, beauty is in the eye of the beholder.

Nevertheless, personal subjective reality does not and cannot negate the existence of a background reality of energy undulations in the zero-point field governed by natural physical law. That background reality is beyond what nature has equipped conscious beings to apprehend and is out of the reach of their sophisticated tools of inquiry.

Einstein could have been assured that the moon was still there even when he was not looking at it. Its waves of probability are forever part of and one with the all-embracing boundless sea of energy. No act of observation can ever snatch those waves out of the moon's energy vortex that is eternally wedded to the infinite universal sea of quantum energy fluctuations.

One's next-door neighbor does not steal the moon by being the first to collapse its wave function resonantly specific to her. The beautiful moon's vibrations get enmeshed with hers and hers with them, and the moon obtains vibrational information from her as she likewise gets from it—forever. When someone sees the moon shortly after she does and its wave function collapses in his microtubules, he, too, gets enmeshed with it as it does with him, and he is enmeshed with her for being intertwined with the same moon. The three of them are unified within the vibrational standing wave of infinite energy in a oneness that unites their spatial and temporal separateness.

INTERFERING WAVES

A quote from the Ashtavakra Gita says: "How wonderful it is that in the infinite ocean of myself the waves of living beings arise, collide, play and disappear, in accordance with their nature." How do waves of living beings collide and play together? It has to do with the intrinsic constitution of waves. The Huygens principle was an idea presented three hundred years ago by Christiaan Huygens, who found that many individual waves coming from different distant sources, called out-waves, formed a single wave front (known as the Huygens combination wave front) coming toward an individual's point of reference in a spherical fashion as in-waves. This leads to a sharing of waves between an entity's own out-waves and the incoming Huygens combination in-waves that sometimes culminates in harmonic wave interference, such that there is no differentiation between the in-waves and out-waves in their unified existence as a standing wave. It may then be said that every single particle, in its dual existence as a particle and a wave, cannot escape the clutches of a Huygens combination wave front. The colliding resonant particle and Huygens wave front are joined as one and arise as a standing wave of living energy.

The same thing can be said of composites of particles. All living and non-living things in the cosmos are united in the standing waves formed from their distinct spherical out-waves and in-waves. The sharing of waves between every finite bit of matter in the universe unites them as a whole; their shared information stored in a fractal holographic pattern. One may recall Thomas Young's double-slit light interference pattern, which is just a two-dimensional way of looking at the picture.

Logically, then, innumerable standing waves have to exist. All crystallized matter, from the atom to the rose, from Fido to his master, and from the sun to clusters of galaxies, are each a distinct combination of waves that acquire infinite variations of patterns and strengths in an eddy, each a standing wave, a singularity of a vortex of tightly resonant frequencies—and all within the one all-encompassing standing wave of the zero-point energy field.

FREQUENCY PERCEPTION

According to American physicist Steven Weinberg in his 1977 book, *The First Three Minutes*, "in our universe we are tuned into the frequency that corresponds to physical [material] reality. But there are an infinite number of parallel realities coexisting with us in the same room, although we cannot tune into them." This statement is made concrete when one stops to consider that the ordinary radio (or TV) station stops playing Chopin when we switch the dial to another station that is broadcasting heavy metal. One no longer hears Chopin because, although the classical station is still actively broadcasting Chopin over the airwaves, the radio is no longer tuned to receive the classical station's frequency range. There are frequencies that are not seen, heard, or felt. There are realities outside our own that our receiving apparatuses are not tuned into. There are exceptions, though. A few differently-wired individuals are able to tune in to those otherworldly melodies.

To some people's minds, it may be inconsequential that what they observe in their surroundings is actually limited only to the vibrational frequency ranges of visible light (about 430–770 terahertz, or 430–770 trillion hertz), which corresponds to 390–

700 nanometers (or 390–700-billionths of a meter) in wavelength; audible sound (about 20–20,000 hertz); and palpable oscillations. We do not hear vibrations below 20 hertz or above 20,000 hertz (the lowest and highest frequencies of audible sound), nor do we see light that is below 430 terahertz or above 770 terahertz (the lowest and highest frequencies of visible light). Our ability to see or hear is limited to narrow ranges.

We take for granted an incredibly immense perception deficit in our everyday lives. Other animal forms and plants utilize some of these vibrations, as discussed in an earlier chapter. The capability of the human body's sensory organs is completely reflective of what nature has provided and what has evolved. No one can hear Perseus's song in B flat or the relic of the birth moans of the universe; nor does anybody see the CMB. Nevertheless, high-precision instruments have presented remarkable evidence of these phenomena.

PERCEPTION AND REALITY

The American transcendentalist and poet Ralph Waldo Emerson (1803–82) wrote that "every particular in nature, a leaf, a drop, a crystal, a moment of time is related to the whole, and partakes of the perfection of the whole. Each particle is a microcosm, and faithfully renders the likeness of the world."

Emerson's statement in some ways evokes the meaning of the Zen word *koan*, from the Japanese *ko*, meaning public, and *an*, meaning matter for thought. *Koans* are riddles to be meditated upon by Zen students. The *koan* presents a paradox aimed at the student's enlightenment on reality while simultaneously showing that logical deductions are inadequate for the purpose. Dogen Zenji, the thirteenth-century Zen master who brought the Soto school of Zen Buddhism from China to Japan, defined *koan* as "ultimate equality," or "sameness," the alikeness in essence of everything that we experience as part and parcel of reality. It is by this sameness in essence that Emerson's leaf faithfully lays before him the likeness of his entire world.

The paradox is not in the *koan* but in a student's concepts while meditating on it. Inasmuch as concepts are defining and limiting constructs of the mind, no idea or thought can ever totally capture and encapsulate the essence of reality, because a thought itself is one with the reality contemplated upon. Reality can only be directly experienced right here, right now. Life is lived by being present in the moment—not thinking about what happened yesterday, nor waiting for something to happen tomorrow.

Everything—all that we see, hear, and touch right here, right now—is in the totality, existing as specific parts while completely enmeshed and united with the whole, and cannot be separated. Individual material realities have differences, each existing in its own sphere even while fully incorporated into and belonging to the same totality. Each personal reality, although vibrationally shared with the whole, intimately belongs to the perceiver alone.

We accept that we may never know how the little old ant of Sinatra's song experiences the fall of the rubber tree plant because, though we all share the same oneness, the ant's frequency domain is just a tad removed from that of a person's in ordinary circumstances. Yet it is bothersome that science in general and physics in particular have not advanced an adequate explanation for experiences of oneness in not-so-ordinary circumstances, that are part of the reality of people in varying cultures and creeds.

This shortcoming is a source of frustration for many scientists, a bunch of whom resist or conveniently discount those anecdotes of uncommon and outwardly mysterious personal experiences. Nonetheless, some of them have chosen to explore solutions outside the box, no matter how uncomfortable their situations become when faced with the all-too-real threat of job loss and fall from prestige. Those who remain in their comfort zones dearly hope for future discoveries to shed light on the problem, or they simply ignore the issue on their merry way to other much more agreeable and rewarding pursuits. Certainly, there is no shortage of riveting, intellectually stimulating, and emotionally fulfilling fields of study in physics and industry that are calling for their attention, effort, and time.

Bohm was one of a few prominent physicists, including Paul Dirac, Alfred Einstein, Alfred Landé, Erwin Schrödinger, John Wheeler, Michael Green, Roger Penrose, and Eugene Wigner, to name just a few, who answered the siren call to unravel the secrets of nature and to seek understanding of its unexplained phenomena outside the confines of convention. Like Einstein with his EPR paper and its hidden variables, Bohm could not accept the randomness, indeterminism, and lack of causality in the quantum world. He believed that there had to be an underlying deepest layer of truth supporting all of visible, audible, and tangible materiality. In his heart, Bohm knew that a hidden mystery lurked behind the ability of the quantum potential to allow instantaneous interaction between individual entities within a totality.

The motion of one particle is dependent on the positions of all other particles in the universal wave function in which every particle belongs, and the motion is subject to the guiding equation of this one wave function. This all-pervading singular wave function evolves according to the Schrödinger equation and never collapses, although specific wave functions of particles within the singular wave function do collapse in accordance with the guiding equation, de Broglie's pilot wave. In this manner, Bohm's subtle guiding force injects causality into the quantum realm. This guiding force is non-material and, therefore, spiritual.

According to Bohm, the causality interpretation of quantum theory "opens the door for the creative operation of underlying, and yet subtler, levels of reality." The presence of a hidden quantum cause allows for the observer effect and entanglement across space and time. When the lover saw the beautiful moon and he was passionately moved to sing, his wave function evolved, as did his beloved's when she heard his tender serenade, as did the wave functions of the birds, which seemed to hush to listen. So, too, did the wave functions of the branches of the trees, which felt the weight of the birds that had stopped and stayed, entangled in the poignancy of the moment.

OF MANY WORLDS AND PARALLEL UNIVERSES

American physicist Hugh Everett III's "many worlds" hypothesis is all about waves and their wave function evolution, which proceeds through eternity without the need for particles to emerge from the collapse of a wave function. Under the supervision of John Wheeler, his mentor, he proposed that although we can observe a particle in only a single position, it actually occupies all the positions allowed it by quantum theory—in different universes. This would mean, therefore, that all kinds of individual realities exist, in the umpteen different parallel universes allowed by the equations of quantum physics.

The possibilities are endless and pointless. That's right. After all, if one thinks that there are really no actual material universes out there and that realities are resonantly forged in the human brain alone, then parallel universes do not exist unless someone is able to tune in to various other rhythms of waves of potentialities and experience different realities, as could possibly be the case in multiple personality disorder.

As far as I am concerned, I can leave parallel universes to waste by the wayside. For all practical purposes, my reality is what I make right here, not in other worlds. And I have the theoretical backing of the renowned English physicist and author Stephen Hawking in his idea that our universe has a wave function that is spread out over all possible universes, just like the quantum electron probability-wave function is spread out all over space. This particular universe is here because here is where its wave function's energy density is greatest for its manifestation into reality. I, as a unique wave function, am in this eternal universal wave function, unified with it as it is unified with me, formless and unmanifest, until my composite frequency achieves resonance with the microtubules of a conscious being and my wave function collapses as the material me in this material universe. My vortex of energy does not crystallize and manifest in any other universe, where the energy density of my probability waves would be lower than in this beautiful universe I live in.

EDDIES OF INDIVIDUAL REALITIES

David Bohm's hypothesis on the emergence of reality says that all objects and events in the "unfolded" visible universe, though relatively stable and seemingly autonomous, are indeed merely "subtotalities" emanating from the unbroken wholeness of the implicate order. These subtotalities are part of and one with the all-encompassing grand vortex of energy, which is like a stream, the surface of which exhibits a changing array of patterns of flow, be they waves, ripples, eddies, or whirlpools, that come and go. But these flow subtotalities of the implicate order do not have an independent and absolute existence. As in the flowing stream, they proceed from the wholeness of the implicate order and are only temporary, forming and disappearing back into the whole, much in the same way eddies return to the main stream that begot them in the first place as a result of the water's flowing movement across plains, rocks, and bends.

According to Bohm, just as the quantum potential guides an electron to follow a precise path, the potential guides the whole process of the unfoldment of totalities in the implicate order into manifest reality in the explicate order. Einstein rejected this idea.

Bohm's hidden quantum potential had too much "spooky action at a distance," even when rejecting the Copenhagen interpretation, Einstein's EPR paper itself invoked hidden variables. Einstein and Pauli informed Bohm that de Broglie had had this interpretation way back in 1927, of which Bohm was totally unaware.

De Broglie had postulated that the waves in the Schrödinger probability wave equations guide or pilot the particles, which are themselves singularities of the waves. He had to abandon the idea of particle singularities because of mathematical difficulties, even though de Broglie thought the mathematics of singularities could not be ignored. He found himself in a "black hole" he had no way out of, and so he did not pursue particle singularities any further.

He found it expedient at the time to revert and accept Bohr's Copenhagen interpretation, complete with all its uncertainties. At the 1927 Solvay Conference, de Broglie presented his matter-particle-wave equations with only "pilot waves" and without the problematic particle singularities.

To many, the introduction of causality and thus determinism into quantum physics pushed the field into the discipline of philosophy and metaphysics, an area of scrutiny inconsistent with the tenets of physical science. To some physical scientists, science must stay within the boundaries of scientific study. Regardless, the idea of causality attracted a group of distinguished supporters, among them Albert Einstein, Louis de Broglie, Guido Beck, and Jean-Pierre Vigier.

Bohm completed his postgraduate studies in physics at the University of California, Berkeley, while working at its Lawrence Radiation Laboratory. It was here where he observed something peculiar in the behavior of the electrons present in plasma, a gas containing electrons and positive ions in high concentrations. He noted that when the electrons became densely packed, the plasma seemed to take on a life of its own, as if the electrons stopped being just individual particles bumping into one another in a random manner. There was rhythm in their movement, and they acted as if they were part of a coherently orchestrated whole. He called his observed phenomenon "plasma oscillation." This unusual experience was repeated at Princeton, where he took the position of assistant professor in 1947. Bohm continued his research on electrons, but this time with metals. Again, he noticed that the normally chaotic and haphazard behavior of individual electrons produced effects as if they were cooperating with one another in a synchronized, orchestrated manner.

In 1959, already in London in self-imposed exile as a result of the McCarthy era of anticommunist hysteria in the United States, Bohm and his young research student, Yakir Aharonov, witnessed quantum connectedness in action in the lab. They observed that electrons seemed to be able to "feel" and were affected by a nearby magnetic field, even though they detected no measurable magnetic flow around them. This was a surprising experimental display of the subtle underlying guidance from Bohm's quantum potential.

Now known as the A-B (Aharonov-Bohm) effect and accepted as a valid quantum phenomenon, the notion initially met strong resistance. Bohm was undaunted. Being the maverick that he was, he loved to buck orthodox scientific ideas. In 1986, the existence of the A-B effect was verified through the sophisticated technology of electron holography.

THE FULL EMPTY SPACE

The atom as a whole is typically described as being roughly 100,000 times larger than the nucleus, which can vary according to atomic weight. For example, a hydrogen atom is 125,000 times larger than its nucleus, and the nucleus in turn is 2,000 times larger than the electron. There are no solid particles in between the nucleus and the electrons and in between the electrons themselves. This space has been described as empty, a vacuum.

When you dig deep into the heart of matter, there is no matter. The vast majority of the atom is empty. In the core of the atom, no single, indivisible fundamental solid building block of matter is to be found. The supposedly tightly bound quarks of all flavors, spin, and strangeness that make up the protons are not there. The reality of an electron itself is fuzzy, existing only as a cloud of energy densities that harmonically flit here and there in waves of probabilities, all spread out across eternity.

In the early seventeenth century, Descartes proposed that the universe is a "plenum" full of matter in which no vacuum could exist; in other words, it was not an empty space. In this plenum, all motion is circular, a trait predicated on indefinite divisibility of particles. Gottfried Leibniz pushed for a more radical idea. In his rejection of the belief in an irreducible atom (called atomism), he claimed that all matter is divided to infinity and is everywhere infinitely in circular motion.

At the zero point, the theoretical absolute zero temperature, all known particles' energetic activities come to a halt, and energy is expected to be zero. Scientists have described this as a quantum vacuum, an empty space. But, in a bizarre way, this zero-point field, where everything is at equilibrium and in absolute stillness, is not empty: the field holds an energy of infinite measure and infinite potentialities. The field is a quantum "foam" of dynamic oscillations of energy in an incessant ritual of manifestation and dissolution. The zero-point field is where energy vibrations are infinitely fine, frequencies infinitely fast, and where all forces—nature's four identified fundamental forces of electromagnetic, gravitational, strong, and weak nuclear—are unified and cannot be individually defined.

In the zero-point energy field, space and time do not exist. The equations of relativity yield black holes of zero size and infinite density, and likewise, the equations of quantum mechanics generate zeros and infinities.

We have established in chapter 7 that any discussion regarding frequency in this book will mean the frequency of light and sound, perceived or not. The infinite energy of the zero-point field infinitely radiates infinite light. Since this energy is all-pervading, light is all-pervading. This infinite, all-encompassing light of the entangled frequencies of waves of probability and of infinite potentialities is therefore present anywhere and everywhere, anytime and all the time, in a never-ending movement of creation and dissolution of specific eddies of energy undulations.

Because light does not rest, it does not acquire mass. Because it is everywhere and always present, it does not have to travel and therefore, has no speed. It accounts for instantaneous action at a distance because light is simply just there, in no time, all the time. Only when light achieves resonance with a conscious observer that such aspects as mass, motion, space, time, and speed are ascribed to it—but only in the observer's mind. Infinite light is the sole source and the sole agent of our material manifestation: the what and the how of our experienced reality.

The effect of zero-point energy on the perceived material world is definitely not zero. Its endless radiation pervades all things and is greater than the energy contained in all of the matter in the cosmos, the earth and all heavenly bodies and space included. This endless radiation must be the truth behind the concepts of Descartes's plenum, Newton's and Huygens's pervading ether, and Maxwell's luminiferous medium that Michelson and Morley never found.

The Dutch physicist Hendrik Casimir (1909–2000) predicted this non-zero vacuum in 1948, when he observed that two metal plates came together when placed parallel to each other in a vacuum, since known as the Casimir effect. A vacuum is supposed to be empty, and there should not have been sufficient pressure to push the metal plates closer to each other. This phenomenon is now attributed to frequency and wavelength. The narrow space between the plates in the experiment excluded wavelengths that were longer than their separation distance. This exclusion decreased the amount of energy in between the plates relative to the ambient energy outside the plates. The closer together the plates were, the more wavelengths were excluded and the greater the outside pressure became.

In the classical world, we can compare the Casimir effect with a closed, half-empty plastic bottle of water that looks fine in an airplane while still airborne but appears crumpled upon landing. Way up high in the sky, the atmospheric pressure was uniformly low both inside and outside the bottle. Now, at sea level, the outside pressure is greater than the pressure inside the closed bottle, so the outside pressure squeezes the plastic in.

That is roughly how the Casimir effect works in the "vacuum" full of energy. No matter that the discernible energy between Casimir's plates is almost nothing, it is not at all zero. On the contrary, each individual oscillation of the zero-point field has, by nature, infinite energy of infinite wavelength and frequency. This notion is beyond the limited perceptive capability of the human apparatus, no matter how absolutely amazing it already is. And our scientific instruments are just as limited in capability.

Early quantum equations had predicted infinite energy density in the empty space in Casimir's experiment. Since particles were the order of the day, scientists set aside the vacuum's energy with the tough problem of infinity and did not see it as a field for further inquiry. Besides, at the Planck length of 10^{-33} centimeters or less, no equation could make the mathematics of quantum mechanics compatible with gravity and general relativity. The laws of classical physics didn't apply either, and time and space were non-existent. This made no sense to the scientists.

Most physicists were not the least interested in tackling that problem or even in raising the question of why it was so. Thus, in the beginning, they summarily ignored all this discovery of eternal bouncing, dancing, and entangling of infinite energy potentialities in the vast expanse of supposed space for the sake of expediency. Nobody seemed earnestly curious about these discoveries, since almost everyone was much too busy being quite enamored with particles and all too willingly being seduced by the overwhelming and heady successes of quantum mechanics.

THE UNIVERSE ON A STRING

The physics community's general disregard for quantum jitters would change with time. The incompatibility between the two pillars of modern physics—general relativity and quantum mechanics at the Planck scale—bothered an increasing number of physicists, and the effort to unite and meld them together finally took hold. Besides, the burgeoning number of particles being discovered added to scientists' uneasiness. One simply cannot have a basic, fundamental theory that allows a hodgepodge mixture of particles and antiparticles in their respective fields, one specific field for each and every particle: the electromagnetic field for the electron and photon, positron field, quark field, Higgs boson field, gluon field, W and Z boson fields, neutrino field, and so on and so forth. The notion of one grand field of infinite and eternal energy undulations creating uniquely patterned individual vortices of vibration makes better sense than discrete particles, each in its very own required field.

Gratefully, even before one could wrap one's head around the crowded zoo of particles in their own fields, the particle approach to reality lost some of its allure to string theory, which is a simpler idea to grasp. To accept superstring or string theory, one has to let go of the mental picture of point particles as the central and basic ingredient of the material world and instead conceptualize a super-tiny string as a stretched filament of energy with the freedom of to-and-fro movement and oscillation in one dimension. Several physicists described a teeny-weeny one-dimensional string inside multiple (10–11) curled-up space-time dimensions vibrating to the tune of an infinite number of simple harmonic oscillators.

In a literal way, Pythagoras was the first actual researcher and practitioner of the "string theory." He progressively halved the length of his lyre strings to compose notes in pleasing harmonic intervals, a fact of scientific importance to mathematics and music and, in turn, geometry. He and his disciples believed that the heavenly bodies moved according to mathematical progressions of musical harmonics, creating a symphony he called the "harmony of the spheres." They sang and made music at the beginning and at the end of their day to attune themselves with the universe. They played music to the sick to heal their body and soul.

String theory hit a resonant chord when, in 1968, Gabriele Veneziano, a young Italian physicist at CERN, advanced the dual-resonance theory to explain the strong interaction of nuclear components: the quarks and the protons and neutrons that they constitute. The excitement quickened when the then five existing string theories were incorporated into one, called M theory, which could easily connote "matrix," "magic," "mystery," or, if you are feeling really bummed out, "murky." Take your pick. The name is not a favorite of mine, either—intuitively, that is, since I am no physicist.

To a lot of scientists, the enchanting call of M theory was its integral potential to be incorporated into any physical theory and law there is, and the idea promised the unequivocal delivery of Einstein's exasperatingly elusive grand unified theory. However, while superstring theory was a step in the right direction, it has failed in the validation department, since, being in a realm where space and time break down, the ever-so-tiny string can neither be proven nor disproven. The energy required to test the theory is just so great as to be impossible to provide.

16
Science Needs a Paradigm Shift

The day science begins to study non-physical phenomena, it will make more
progress in one decade than in all the previous centuries of its existence.
To understand the true nature of the universe, one must think
in terms of energy, frequency and vibration.
—Nikola Tesla

At this point in particle physics research, a lot is already known about electrons and the atomic nucleus. The quarks that make up protons and neutrons remain unclarified, however, and their properties, including their vibrational frequencies, have not been defined. Their attraction to one another is so strong that scientists, so far, have found it impossible to wrest them apart to allow for individual scrutiny. This is a mystery. Or is it, really? The question demands a closer look.

The premise that everything vibrates seems an appropriate point to start. According to quantum field theories, quarks are vibrations in their fields. Their vibrations very well could be of infinitesimally high frequency and in perfect resonance with one another. In consequence, they could be intertwined so tightly that nothing we mortals possess in our laboratories can pull them apart. The same principle could be just as true in the case of the individual protons and neutrons that quarks constitute.

The nucleus itself may only be waves of resonant vibrations instead of the solid central core of the atom from which, as we'll recall, Rutherford's alpha particles bounced back. The alpha particles themselves just as easily could simply be vibrations existing only as clouds of probability waves, much in the same manner that electrons exist. In such a scenario, the alpha particles' frequencies cannot achieve resonance with the nucleus, so they have to harmonically bounce off.

Maybe all those different fields are not necessary at all. There could just be one unified and unifying vibrating field, a grand standing wave of energy holographically storing information within its waves of interference harmonics. All the localized vibrations that scientists have assigned field names to may just be focal energy vortices of harmonic wave interference stored as unique holograms, like eddies in a stream, that materialize as particles or faces or stars, strictly in our consciousness. There would be no need to talk about particles in their own specific force fields, ever.

The signatures of particle collisions, including those indicative of the Higgs boson, could very well be newly created harmonic arrangements resulting from wave entanglement and separation during collisions in installations such as the Large Hadron Collider. They could simply be new eddies of frequency created from the disruption of frequency patterns, which could be why, with each increasingly violent collision in increasingly powerful particle colliders, scientists keep finding more fingerprints of newer particles. Particle discoveries could very well be endless, since there are infinite combinations of frequencies in infinite patterns of entanglement.

E = HV = MC² INSTEAD OF GRAVITY?

Johannes Kepler once said, "Gravity is a mutual affection between cognate bodies toward union or conjunction (similar in kind to the magnetic virtue), so that the earth attracts a stone so much rather than the stone seeks the earth." Gravity is defined as the force that attracts any object with mass toward the center of the earth or toward another object with mass. Since objects of mass vibrate and have specific frequencies, they very well could be attracted to one another through resonance.

Take the principle that governs the laser beam. Monochromatic same-frequency laser radiation, with its waves oscillating in synchrony with one another, is coherent light. The photons are tightly bound, resonantly glued together as one in a standing wave, such that the laser beam's edge is completely devoid of fuzziness. This allows the beam to cut with clean sharpness, as exemplified by the Nd-YAG laser (discussed in chapter 8), which can precision cut a clear white diamond or cleanly remove a clouded posterior capsule of an eye lens.

Since mass and frequency are equivalent, the gravitational attraction between bodies of mass represents attraction between their resonant frequencies. The "force of gravity" is the strength of coherence between their resonant frequencies. The same principle can be applied to inertia. It is so much easier to lift a feather than a stone, since the composition of the feather is less resonant with the earth than the composition of the stone is.

The reasoning offered above reveals tremendous implications. The resonant attraction, the mutual affection between the earth and Kepler's stone, becomes fundamental and inherent, and the equations of general relativity and gravity are only mathematical expressions of the harmonics of that resonant attraction. The same can be said of the human body's gravitational attraction to the earth. The resonant frequencies of the earth (7.83 Hz) and the human body (12.3 Hz) are close enough that we can say that the earth and the human body are resonantly attracted to each other just enough for humans to stay grounded and not fly off into space.

Still, the higher frequency of the human body compared to that of the earth allows us to jump up and away from the ground temporarily and not stay glued to it forever. Monks and master meditators are said to be able to levitate in defiance of gravity through an increase in the composite frequency of the body during the transcendent state, thus decreasing the degree by which they resonate with the earth.

Therefore, we can say that the gravitational field is just a composite of vortices of harmonically related frequencies. This notion makes sense of centuries of observation that heavenly bodies are spatially harmonically arranged, and it makes sense of the manner in which planets revolve around suns and in which stars revolve around galactic centers. These events exist against a backdrop of likewise harmonically arranged cosmic microwave radiation. Attraction via resonance could very well be Kepler's mysterious unnamed power that he thought emanated directly from the sun.

Everything else follows. The attraction between quarks is no more than vibrational resonance so coherent that the attraction tightly entangles the quarks' vibratory waves so much that they cannot be pulled apart. Furthermore, vibrations with identical frequencies form standing waves that go nowhere, stabilizing the bond. The singularity of the quarks' vortices, as standing waves of coherent resonant frequencies, strongly bind quarks to one another. It is no longer necessary to invoke the separate entity of a strong nuclear force to hold them tightly together.

Taking this reasoning a little bit further leads to the following line of thought. Quantum force-field theory allows infinite energy and infinite frequencies. Infinities are renormalized or altered in physics calculations. If renormalization is not allowed and the equations $E = hv$ and $E = mc^2$ are populated with infinite energy of pure resonant frequency of infinite light, then the mass obtained would be infinite. This statement may seem stupid, but it deserves further attention.

In the realm of material space-time, according to the equation $E = mc^2$, the black hole has a mass equivalent to the mass of the star that died. But the mass is infinitely dense and compacted that it cannot be measured; it has no size. Its gravitational force is so strong that light cannot escape from it. No information about anything travels out of the black hole singularity. Besides, its unbounded infinite energy cannot be assessed or quantified by any apparatus in bounded limited space-time. The equations of general relativity do not apply to the black hole singularity.

In the quantum realm of vibrations, this black hole singularity is the core of the vortex of harmonically arranged probability waves of the galactic spinning disk. In this singularity, frequencies are of such resonance that a stable standing wave is formed, and the oscillations do not travel. In here is the heart of the black hole vortex where everything is still. In this singularity, energy, light, is infinite and eternal.

Since $E = h\nu = mc^2$, both equations, Planck's $E = h\nu$ in the quantum world and Einstein's $E = mc^2$ in space-time, serve up a black hole and its singularity of infinities. Quantum mechanics and general relativity are not at odds with each other. They are different expressions of the same thing—just as Schrödinger's wave function equations describe waves and Heisenberg's matrix-mechanics equations describe particles, all the while dealing with the same exact entity.

Quantum mechanics and general relativity are at odds, though, in the brain microtubules, where electromagnetic reception has to necessarily conform to the limitations of the material brain apparatus, which receives and integrates only a narrow range of electromagnetic wave frequencies; limitations that restrict mass to be understood only as a space-time construct of our mind.

Our understanding of the quantum is hindered by the inadequacy of our ordinary personal consciousness, which cannot comprehend or imagine anything as small as a Planck length, much less infinities and thus, renders infinities, zeros, and nothingness incomprehensible; a circumstance that is overcome, narratively, during a state of spiritual bliss in prayer and during a transcendent state in deep meditation.

The incompatibility between the two pillars of physics is a mental construct. Still, we have to be forgiving and accepting of our imperfection and be compassionate with our fellow human beings. Space-time is our only and most lovingly cherished reality, our intimately personal perception of unfathomable transcendent zero-point field energy.

This is the crossroad where the terms "force of gravity" and "gravitational field" may have to be redefined, or thrown out altogether, and more appropriate terms coined. This is where the current understanding of Einstein's general theory of relativity has to be reinterpreted or given a new function: not as law and the final description of nature but as an expression of the underlying basic law of vibration and resonance.

There has to be a deeper physics that incorporates both the quantum realm and general relativity's warped four-dimensional space-time world. This hidden physics might just be awaiting scrutiny through vibrations, frequency, harmonics, and resonance; through the scores of new, unfamiliar, and enchanting exploratory music of particle pings, fractals, and beat frequencies.

E = HV INSTEAD OF SPACE?

Immanuel Kant said that "space is not something objective and real, nor a substance, nor an accident, nor a relation; instead, it is subjective and ideal, and originates from the mind's nature in accord with a stable law as a scheme, as it were, for coordinating everything sensed externally." Kant could have made that statement only yesterday, even though he proposed that philosophy in the eighteenth century. About a century earlier, Leibniz had advanced the idea that space is only a geometric relationship between objects in terms of their distances and their positions relative to each other: a relationship that exists only in the imagination. Geometry is dictated by frequency and number: that is, resonance and harmonics. Space, therefore, is created in the mind by the separation of objects according to the geometry of their frequency harmonics.

If heavenly bodies are arranged according to harmonic intervals and electron orbitals are sequenced in the same way, then the distance between someone and the moon and the space between her eyebrows must be a matter of harmonic arrangement of frequencies as well, resonantly grasped and brought to life by the conscious mind. It is this mental geometric order that we call space. There is no space existing independently outside of the mind, and there are no material objects out there to define their' boundaries. Space is an idea, a mental representation of the boundless fundamental of energy vibration.

Leibniz did not stop with space in rejecting absolute existence. To him, time is the mathematically sequenced appearances of objects as understood by the mind. As with space, time is an "abstract structure of relations" in which "actual and even possible bodies" are located and move. There is no independent absolute time. Leibniz's notion was right on, even though physicists during his era had no knowledge of the microtubular quantum resonant processes that determine the human experience of space-time.

Nikolai Kozyrev, a Russian astrophysicist who suffered the horrors of concentration camps at the hands of Stalinist Russia, spent a lot of effort studying and developing his concept of the "flow of time," about which he wrote numerous scientific papers in Russian. One experiment of his involved astronomy. He studied the energy flow from stars with a special telescope designed to detect "time flow," or energy flow independent of measurable electromagnetic flow. With light and known electromagnetic radiation blocked off, his instrument detected energy flow from a star, most prominently at its calculated true position. Curiously, his instrument also detected energy flow, albeit weaker, from the star's past position and also from the site of its future position.

What is the meaning of having energy flowing simultaneously from the past, present, and future? And what form of energy would be coming from a star that is not related to its measurable electromagnetic radiation? Is this phenomenon the macroscopic equivalent of Schrödinger's quantum probability waves, which are spread out all over infinite space but still send unseen energy frequencies from higher-density areas with greater concentrations at the star's true position?

If that's the case, then the transfer of information from the three positions would be instantaneous, all done at the same time. The present position would not follow the past, and the future position would not come after the present. The present true position would have greater energy flow because its energy density would be amplified by numerous observer's energy in the many "now" moments of its detection in space-time.

The frequencies that are past, present, and future are all omnipresent in their clouds of probability wave functions, and it is the unfolding of information that makes time linear. This proceeds through harmonic evolution in accordance to Bohm's explicate order, implemented by microtubular resonance frequency bands in fractal arrangement—from the tiny tubulins to the entire brain. Each frequency band, quantum-mechanically acting like a clock, orders the harmonic evolution of conscious experience.

We must reconsider Newton's concept of duration, his original idea of true, absolute, and mathematical time that flows steadily and unerringly forever, unaffected by anything external. In light of what we have discussed, we can now accept his duration or true time to march on uninterrupted, through the mathematical progression of harmonics in the constantly changing dance of energy in the quantum foam that underlies our reality. True time mathematically flows steadily through the evolution of each and every complex entanglement of frequencies, wave interferences and variations in their mathematical expressions, consistent with the fractal holographic enfoldment of information into Bohm's implicate order.

We experience common or relative time (the external measure of Newton's true time) that rules the domain of clocks and motion, through human consciousness as a series of bursts of awareness, the occasions and events of consciousness upon the collapse of the Schrödinger probability wave function. This flow of unfoldment of frequency harmonics in the conscious mind is then designated numbers as hours in a day, days in a year, and years in a lifetime and is decorated with faces and places, highlighted by celebrations and happenings, punctuated by births and deaths of loved ones, and numbered by rings in a tree trunk.

Time is only a relative measure of the flow of the quantum unfolding of frequency harmonics that is presented to the brain for resonant incorporation into human consciousness, proceeding in a comfortable sequence of bursts of awareness of the materialized energy potentialities in the mind. Common or relative time as we know it is purely a mental event.

EINSTEIN'S RELATIVITIES GO THE WAY OF GRAVITY, SPACE, AND TIME

The restriction by special and general relativity that nothing can travel faster than the speed of light in gravity-warped space-time exists because electromagnetic frequency alone is what the brain resonantly receives and processes. As apprehended by the brain, light waves from a source interfere with light waves from other sources with which they get entangled (such as air, water, or objects) and are slowed down or bent by an energy-dense object like our sun. That was what happened to the light from the Hyades

249

star cluster when the moon totally eclipsed the sun to let Hyades's light through and exceptionally allow its brilliance to shine on Albert Einstein's gem of an accomplishment.

There is no possibility for observed light, highly entangled with some other light waves, to reach the measured speed of light in a vacuum. Speed is simply the conscious mind's perception of the evolution of harmonic entanglement of electromagnetic radiation from the zero-point field. Space is the mind's representation of the degree of harmonic separation between the energy vortices that beget observers and objects surrounding them. The distance between perceived objects of mass, between the bright sun and the pale moon, and between the comfortable sofa and the pesky ringing telephone is only a mental impression of the harmonic arrangement of their underlying vibrations.

Space and time simply are constructs of the brain's resonant microtubular apparatus, and 186,000 miles / 300,000 kilometers per second is a speed limit imposed by the brain alone. Light, the electromagnetic radiation of the zero-point field that microtubules accept and process as information, is already present here, there, and everywhere. There is no actual space that it has to traverse in so much time, and there is no real time by which it has to finish traveling. Light has no speed.

Assigning gravity, space, and time to the all-pervading field of vibrations resonantly discerned by our conscious minds makes Einstein's relativity theories purely mathematical and geometrical expressions of frequency. There is no actual solid ball that rolls on a trampoline according to the warping of its fabric caused by a heavier ball and there are no actual solid planets revolving around a star according to the gravitational warping of the framework of space-time. Einstein's moon is not a huge clump of matter in the sky. Objects are only resonant harmonic vortices of vibrations that come to life as apparent solids in a specific place at a specific time as interpreted by the microtubules of the brain.

That there is no real external world of sticks and stones and events out there makes sense of Galileo's principle of relativity. There is no enclosed cabin in a uniformly floating ship, and there are no solid objects called flies, butterflies, or fish in it. There is no ball being thrown toward the bow or stern, no person jumping up and down in an enclosed cabin out there. There are only vibrations and their wave functions resonantly integrated in the mind of the observer, where objects in space-time are experienced in bursts of awareness upon the collapse of their wave functions.

Because resonance and harmonics orchestrate the "now" moments of one's reality in the enclosed cabin, the dripping water never misses the bowl, jumping people always land where they are supposed to land, and the flies and butterflies flutter around with normal speed to all sides of the cabin as if the ship were still moored motionless on the dock. The whole scene is a series of perfectly harmonically organized "now" moments in the mind.

The same holds true in a steady jumbo jet with the windows closed. But if we look out and include the probability waves of fluffy clouds and distant mountains with the probability waves of the airplane cabin, then the mathematics of wave frequencies and their harmonic separations will have to apply, and our impression will be that of movement away from the clouds and mountains.

In the world of vibrations, resonance, and harmonics, all of the preceding scenarios make perfect sense.

Electromagnetic waves and their interference patterns are what the brain microtubules receive, integrate, and interpret as the distance between trees and the motion of their branches, the captivating cadence of poetry, and the melancholy wail of a flute floating in the lonely darkness of night: everything in our reality.

The frequency of light waves is what implements the resonant transmission and reception of information by the microtubular mechanism. Light is the one and only source and the sole agent of manifestation of thought, matter, motion, common time, and all their subjective attributes—the "qualia" of human consciousness.[12]

Where would this discussion leave Einstein's equations of general and special relativity theories? Well, they stay as valid expressions of vibration as apprehended by human consciousness. The equations of Einstein's general and special relativity reflect the entanglement of the never-ending quantum dance of energy with conscious human experience. But new words have to be invented to fully describe what the equations reveal. The "warping of space around massive bodies" is a lame description of reality that does not even tantalize with a glimpse of the essence of resonant entanglement on a grand universal scale, much less capture it.

ENTANGLED FREQUENCIES

Frequency transmission and resonant reception are the way of nature. They account for the ability of plants to tell neighbors from family members, for whom they sacrifice themselves in times of danger in what appear to be intentional acts of altruism. Plants respond to music, human voice and attention and are intertwined with their growers' emotions. Dolphins sing, DNA proteins oscillate in harmony. The CMB, the ancient relic of the first light of the universe, is sound frequency that birthed this musical universe.

Resonance accounts for the strong bond between lovers and between all human beings to different degrees. No one would dare defy the strong emotional connection between Fido and his

---

[12] Qualia are subjective perceived sensations, feelings, or experiences that are difficult to put into words, such as agonizing pain, tender motherly love, or the sinister inky blueness of the ocean.

master. Beyond words, they understand each other's vibes. Resonant frequencies could be the reason why identical twins, partners in life, the lie-detector expert and his plants, twin stars, galaxies and their neighbors, and suns and their planets and moons are fast-entwined and forever united.

We might feel close to our family members because our shared tribal vibrations harmonize better and bind us more intimately than with others outside our family. It could be that one can tell that a person belongs to a certain family from her particular pair of eyes because they, like the eyes of other family members, were formed to the tune of the same melody. It could be that we feel an instant rapport with a perfect stranger because he or she has the "same vibes" stemming from similar likes, dislikes, ways of thinking, and moral inclinations. We might become mesmerized and feel more affected by a certain topic of discussion because we resonate with it. We might also feel more empathy toward people who are dealing with certain life difficulties because we are also in vibrationally similar situations.

This leads us to think of what Matthew 25:40 (KJV) reports Jesus to have said: "Verily I say unto you, Inasmuch as ye have done it unto one of the least of these my brethren, ye have done it unto me." This biblical passage expresses the principle of interconnectedness through vibration between the divine and all living and nonliving inhabitants of the earth, including the whole universe, so that whatever we do to anyone or to anything at all, we do so to God and to our own selves as well.

Many of us have experienced the feeling of discomfort when an instrument in a junior symphony orchestra is out of tune—how we just cringe at the dissonance and how we wish for the sound to stop. It follows that we often feel hatred and anger toward the more violent members of the human race because such people have dissonant vibrations, and we wish to be rid of them, parallel to the bouncing of Rutherford's alpha particles from the atomic nucleus and the explosion of a supernova from core energetic rebound in space-time.

This scenario brings to mind the notion of sin, the spiritual "separation from God" that is taken very seriously in spiritual teachings. Sin could be one of vibrational and harmonic separation from the fundamental frequency of the zero-point field of omnipresent energy—from the light, the *logos*, the first sound that marked the beginning of perceived time.

All the previously described phenomena in the quantum and the macro territories are accessible to inquiry from the perspective of frequency and harmonics, probably better than from the perspective of solid particles and the indirect evidence of their collisions at the gargantuan LHC and similarly built laboratories. Unfortunately, most of the physics community, possibly excepting a few who are involved in biomolecular and quantum chemistry, have focused on particles while we, the people, spend oodles of money on them.

It is high time to start thinking beyond particles and strings. It is time to lay aside the gluon, the weak nuclear forces, the Higgs boson, and the graviton. It is time to reevaluate entrenched ideas and descriptions of reality. It is time to coin new terms and invent ways of expression and elucidation to reflect the long-needed recognition of their vibratory origin.

We need a shift in the mode of investigation—from matter particles to frequency harmonics, as Tesla recommended, and to distinct vortices and standing waves within the one and only standing wave of a universal symphony that begets you and me, musical offspring of the stars. The totality of the quantum and the gross—that is, the whole of our cherished reality—is all in the music. Frequency is the one reality from which everything emanates and it deserves, nay, commands, very serious inquiry.

There is only energy vibration, frequency, and resonance. In his book, *The World Is Sound: Nada Brahma: Music and the Landscape of Consciousness*, German jazz musician and author Joachim-Ernst Berendt (1922–2000) said, "at the root of all power and motion, there is music and rhythm, the play of patterned frequencies against the matrix of time. We know that every particle in the physical [material] universe takes its characteristics from the pitch and pattern and overtones of its particular frequencies, its singing. Before we make music, music makes us." Foreseeing a revolution in physical science, John Wheeler once remarked, "And when it comes, will we not say to each other, Oh how beautiful and simple it all is! How could we ever have missed it so long?"

Light. *Logos*. Frequency. It is the only one. It is that simple.

WHAT TO DO?

High-tech computations have enabled astronomers to record the moans from the birth of the cosmos from the CMB. In the future, mathematicians and physicists could very well come up with computations of the harmonics of nature to guide the study and understanding of reality. Maybe they could start with the music of the quarks or their beat frequencies, which are low enough to be gadget friendly. Scientists have already sonified particle-collision data, in the form of pings, at the LHC. One can now listen to the Higgs boson make music as it decays.

Stephon Alexander (b. 1971), an American cosmologist, jazz musician, theoretical physicist, and author of *The Jazz of Physics*, espouses the connection between music, physics, and the cosmos. He poses that we can begin to understand these things and life itself through sound, harmony, and resonance. Maybe listening to the music of a thinking brain, a joyful heart, or a person in ardent prayer or in a trance, aided by differently-wired people who can tune in to their frequencies and reproduce them—as the speaker Sharry Edwards (mentioned earlier in the book) did with the frequency of the conference room—is an exploration worth embarking on. Many musicians, especially jazz musicians, have composed elaborate musical pieces by drawing from equations of the quantum theory of waves and Fourier transformations.

The above are just my humble thoughts. Physicists, computer geniuses, mathematicians, other scientists, and musicians are much more able to offer infinitely better ideas than I. Maybe humans are getting closer to arriving at some understanding of the object of Einstein's single-hearted scientific quest. But, maybe we will never ever reach that understanding if the exploration is based exclusively on current scientific experimentations. No longer should scientists be satisfied with theories and computations alone. Science must be coupled with wisdom gained from actual living.

To have only hard scientific doctrines dictate the way we conduct our lives and shape the future of the whole universe without the wisdom gained from the lessons taught by history, and without the mellowing influence of direct human experience is foolhardy. Science without conscience brought us the horrors of Hiroshima and Nagasaki and the dark cloud of a nuclear holocaust that hovers over us.

Most importantly, to think that all we should accept of life is only that which hard scientific inquiry can prove is a crime against truth and personal integrity and goes against people's private and shared mystical experiences of nature and its phenomena.

Even scientists themselves realize that the mathematical pictures they use in their equations to describe nature are only representations that do not precisely define ultimate reality, and as such, they cannot be considered much different from illusions. There is no certainty that the physicists' current quantum-mechanical description is exact and not just an approximate picture of an underlying fundamental law. Scientific facts come up woefully short of being able to impart a full understanding of nature. It behooves us to keep an open mind even against the noisy backdrop of scientists' very strong personal prejudices and deeply rooted tribal values. As Poincaré said, "Science is built up of facts as a house is built of stones; but an accumulation of facts is no more a science than a heap of stones is a house."

Although scientific rationality has a necessary place in our lives, pure rationality is not enough for living. Rational thought certainly counts, but so do selflessness, loving devotion, and the sudden realization of an inexplicable and deep connection to something grander than life that is evoked by moving poetry, witty humor, beautiful music, expressive art, glistening dew on a velvety crimson rose, and silver moonlight shimmering on playful wavelets in some secluded bay.

Science does not have a way of probing, imaging, extracting, sorting, filtering, dividing, titrating, measuring, quantizing, entangling, attaching mathematical symbols to, and creating equations for all those subjective human attributes and the uniquely individual quality of raw perception. Nonetheless, they are very much there. As the ancient and wise Star Wars Jedi master Yoda would probably be inclined to say: "Of our reality, unified part and parcel they are."

There has to be room for both. If we could conduct our lives exclusively according to the dictates of cold reason, then our lives would certainly be so predictable that, in no time, would become downright boring. Conversely, if our course in life could not be designed and planned according to reason and to the results expected from every plan, word, or deed, then everything would have to be left to chance, and there would be no reason at all for setting any goals.

Just like the song sung by the late actress and singer Doris Day, "Que será, será, whatever will be, will be," so will people's lives, heartbreakingly, be left aimlessly blowing in the wind. Gone, too, with the blowing wind, will be that enormously gratifying sense of accomplishment after a job well done. Most assuredly will we lose our moral sense of duty, and just as surely will we shed total responsibility for all our actions. In this woeful scenario, the concept of fulfillment in life would be as a lonesome stranger indeed. We have to strictly hold ourselves accountable. According to the spiritual teacher and psychiatrist and author David R. Hawkins (1927–2012) in his book *Power vs. Force,*

> everything in the universe constantly gives off an energy pattern of a specific frequency which remains for all time and can be read by those who know how. Every word, deed and intention creates a permanent record. Every thought is known and recorded forever. There are no secrets; nothing is hidden, nor can it be. Our spirits stand naked in time for all to see—everyone's life, finally, is accountable to the universe.

It is time to balance the scale between scientific rationality and the wisdom gained from experience. It is time to make the body of scientific knowledge more precise. The quest for knowledge through particle physics has led to a corner jam-packed with particles that just keep on multiplying, muddling the clarity science is seeking. The dense fog of particle physics is clearly hindering our vision of the truth.

The current (and expensive) methods of information gathering are disappointingly inadequate for uncovering the fundamental principle that governs reality. Intuition resonating with absolute truth, light, and love must complement scientific exploration and logical deduction. Wolfgang Pauli said that, "a synthesis embracing both rational understanding and the mystical experience of unity is the mythos, spoken or unspoken, of our present day and age."

It is now imperative that the scientific community and the rest of humanity who are so inclined, in harmonious cooperation with one another, embark on a different path and take a new approach. Listening to the music of nature through its unseen vibrations has already been done with "Mary Had a Little Lamb," Perseus's song, and the CMB. So, maybe it is technologically feasible to listen to the music that exists between people in loving relationships and to tune in to the vibrations of mystics in a trance and differently-wired people during their extraordinary experiences. This path seems more reasonable and appealing than studying particle collisions in ever-larger hadron colliders. The wealth to be uncovered is enormous, the field is vast, and the technology for inquiry is already at hand.

TYING IT ALL UP

I think this work has outlined the necessary components for a fresh theory on reality. When I make an egg roll, which in my country of birth is called lumpia (loom-pyà), I lay out the egg roll wrapper, spread the ingredients of seasoned, finely sliced cooked meat and vegetables over the wrapper surface closest to me, and roll them all into one. I then wet the edge of the wrapper farthest away from me with egg wash to glue the whole roll together. Before serving, I fry the lumpia rolls to a golden brown.

For the scientist's lumpia of reality, the wrapper is the all-pervading immanent light in the zero-point energy field. The ingredients are the quantum waves of probability and the microtubules of the conscious human brain, spiced with the kaleidoscopic entanglement of their frequencies, all glued together by resonance. Just as frying is the last step before serving the lumpia to my hungry family, doing the equations of Mother Nature's harmonics readies the magnificent cosmic lumpia to be offered to humanity.

It is now the zero hour to pay deliberate attention to what has been persistently knocking at scientists' and ordinary people's doors and staring them squarely in the face. Vibration. Frequency. Resonance. Nikola Tesla could very well be proven right. The scientists' search for the grand unified theory or the theory of everything has been going on for years on end: a prolonged, futile search based on particle physics. We have to choose another way, as the beacon that Einstein and other sincere scientists have focused on could be only illusory, a mirage of particles to which they tenaciously cling.

The elusive theory of everything just might easily be the transcendent zero-point field energy frequency that is resonantly unified with the energy of human consciousness and experienced in space-time by the mind. We could simply and honestly call this the grand unified resonance theory.

Everything vibrates, from the zero-point field energy jitters to the whole universe. Everything that vibrates has resonance. We are connected to the vibrations of the zero-point energy field through resonance. Resonance is the only way life on earth and in the whole universe works. A scientific theory that is solidly built on vibrations, frequency, and resonance ends the painful, drawn-out search for the grand unified theory of reality that Einstein devotedly pursued through his final years.

The grand unified resonance theory. It rings true. It is the key to the belief in a transcendent infinite and eternal God.

17
The Power of Love

Ah! Sweet mystery of life
at last I've found thee;
Ah! I know at last the secret of it all;
All the longing, seeking, striving, waiting, yearning,

The burning hopes, the joy and idle tears that fall!
For 'tis love, and love alone, the world is seeking;
And 'tis love, and love alone, that can repay;
'Tis the answer, 'tis the end and all of living,

For it is love alone that rules for aye!
—Rida Johnson Young

Love. God. Pure consciousness. Pure light. Zero-point energy. They could all be just names for one frequency, one tune that begets, embraces, and entangles everything in the universe as one. This singular source has been believed, taught, experienced, and lived by the ancients and present-day sages and is now only scantily unveiled and timidly proposed by today's physical science. Its infinities are out of the reach of ordinary human contemplation and inaccessible to highly scientific postulation and technologically sophisticated laboratory experimentation.

One has to let go of rational thought and scientific rigor to have even just an inkling of its immensity and boundlessness, to which mortal comprehension is no match. Arriving at an intimate knowing that flows from focused attention on the vibration of love is the goal. One does not have to be an experienced meditator, a sage, or a mystic to achieve resonance with love. Kind thoughts, sincere intentions for the well-being of another creature, and loving words and deeds are all harmonics of love.

We are not mere thinking entities. We are also sentient and feeling organisms. We enjoy a privileged state of being that allows us to have our own closely personal experience in addition to those shared with our community, preeminently exemplified by the great emotional shock and pain unexpectedly brought on by the tragic death of the beloved Princess Diana. Her death was all at once intensely personal and widely shared. She was the "people's princess," a brilliant star whose warm and gentle light, which had touched people all over the world, was senselessly extinguished in a car accident in Paris.

Diana's death was an experience marked by a strong emotional resonance that united many people of various cultures in a convergent moment. It was a universal feeling so intense that it registered a major blip in the "random event generators," or REGs (electronic machines that initiate a random stream of the numbers 1 and 0, like an electronic coin flipper), scattered all around the globe. It must have been the momentary collective rise in the intensity of the interrelated electromagnetic signals from the hearts, minds, souls, and bodies of people and nature quantum-physically connected to the princess, that registered the strange but indisputable blip in the REGs in the days of global mourning following the moment of her death. People then were of one heart and mind, resonantly united in grief and heartbreak from the shocking loss of their dearly beloved princess.

The brain holds no exclusive claim to all nervous tissues that both transmit and receive information in the form of electromagnetic vibratory signals between humans and this vast universe. All cells of the body do. The organ with the biggest electromagnetic field of all, even much bigger than that of the brain, is the heart, the nerve-like tissues of which send out vibrations that can be reliably measured and recorded by an electrocardiograph machine, imaged by Kirlian photography, or electromagnetically mapped by the newly researched technique of magnetocardiography. The frequency of the heart's vibrations is entangled with the whole environment and with the all-pervading infinite universal energy, the pure frequency that operates outside the reach of the normal five senses.

The consciousness beyond a human being's externally acquired knowledge is what the child prodigy, the prophet, the remote reader, the savant, and the clairvoyant tap into. It is the quantum holographic source of "knowing" for the mystics, sages, and adepts, whose human consciousness, through meditation or prayer, transcends the limitations of the material state of being and vibrationally enters the zero-point energy field, becoming one with it while staying in manifest material form in the mind of the human observer.

For us ordinary mortals, we only have the heart, which, in the quiet stillness of the mind, just might catch a glimpse of that inscrutable wisdom that is veiled from our eyes. Have you ever had a very strong feeling, a deep knowing in your heart, about a certain something that later proved to be correct, one that you could not explain by using your clever thinking brain, how and why you simply knew? Have you ever had a desire, of which you told no one because it seemed so unattainable, that fell into your lap somehow, as if by magic? If your answer is yes, be glad and give thanks. Your life is all the richer for that. And if we all look for cues in our environment and listen to our hearts more, our lives will be chock-full of those awe-inspiring intuitive moments and astonishing synchronicities.

Still, it would be wise to heed this old advice from Hermes Trismegistus: "Keep your mind ever on the Star, but let your eyes watch over your footsteps, lest you fall into the mire by reason of your upward gaze."

ACTING ON LOVE

Love is charity. Charity is kindness. Kindness is an act of love. Sometimes it is difficult to hold a loving feeling for someone who is revolting to the senses or for someone who has caused us pain in any way. It is easier to coax ourselves to practice kindness even when the loving feeling is not there. Love is action, and the feelings that one may find wanting at the start, will follow in due time. Resonant harmonics will take care of that. The consistent practice of loving deeds entrains and synchronizes the brain, the heart, the whole being, and the universe to the vibration of love. Everything else follows. It is physical law.

Love is the basic frequency that connects all creatures in the universe. What we experience as bad or evil is only dissonant frequency that has to be retuned back to harmony with the energy of love. Being out of tune is not set in stone. Being in and out of harmony with love is part of the dynamic and eternal harmonic evolution of the eddies of probability waves in the unseen world. Trying to stay in tune with love is like the plight of Lord Kelvin's atomic vortex rings and the knots reproduced in water by Kleckner and Irvine, which repeatedly strayed apart and came together again. Vortex rings, knots in water, and creatures of the universe behave no differently from one another. They follow the same laws of physics and abide by the same principles of nature.

Dissonance can be influenced by every vibration it encounters and gets interwoven with. Intention can influence the material manifestation of the waves of potentialities and group intention can do so even more by virtue of energy density, the fundamental principle underlying the space-time notion of critical mass.

Therefore, meditation, fervent prayer, goal setting, and sincere effort toward harmonic reconnection with pure energy for a healthy planet, common good, and enlightenment and evolution of human consciousness will work. Humanity can influence the world, and an earthly life of peace, joy, beauty, harmony, and love woven into an all-inclusive symphony, is ours for the taking. We owe it to ourselves, to future generations, and to the universe as a whole to claim that life of our desiring.

WIELDING THE POWER OF THE SCEPTER OF LOVE

Hermes said, "he who understands the Principle of Vibration, has grasped the sceptre of Power." The participatory creative power of the conscious observer is Hermes's scepter of power through vibration. It is high time that the community of humanity wields the scepter in abiding harmony with the all-pervading pure consciousness, the music of love of the universe.

John 1 verses 4:7–8 (KJV), says: "Beloved, let us love one another: for love is of God; and every one that loveth is born of God, and knoweth God. He that loveth not, knoweth not God; for God is love." If loving everybody is too ambitious and too big an order for the day, we can always start with the simple sincere intention and effort to practice kindness in every thought, word, and deed. As the Holy Bible tells us, a widow's mite (a small coin) wholeheartedly given is worth more than gold many times its weight but packed in a bag of hypocrisy. The universe honors the vibrations of earnest devotion no matter how small. Such is the power of heartfelt desire, sincere intention, and genuine effort in creating one's reality from the infinite potentialities of the sea of constant energy jitters of pure consciousness.

Fighting words and armaments of destruction are in resonance with the violent members of the human race. They amplify the energy of discord, widen the separation between fellow beings, and tune them out and away from love. One has to choose to resolutely cling to love.

Of the over 7.5 billion people of the world today, roughly eight in ten people profess to a belief system. If all spiritual believers acted together, joined by the other members of society who are similarly inclined, the total number of participants would be tremendous. This can only mean that the collective energy density may reach a critical level that effectively changes the frequency of the disharmonious members of the human race to one that resonates with love. If the numerous places of worship and secular meeting places were to be utilized for a unified intention to love, and if the many communication outlets such as social media were likewise engaged, then "calls for love and world peace" could be synchronized to achieve maximal energy density whenever they are held. Because of the different time zones across the globe, people will always be wielding the power of love every minute of the day, three hundred sixty-five days a year.

The above is a challenge to the hierarchy of organized religion and other institutions of the world, to the owners, drivers, and users of social media, and to all people of good will. Humanity holds the scepter of power; why not wield it in unity? It is the responsible thing to do in this tumultuous state of world affairs humanity endures.

It is a worthy aspiration for us to walk through this life in abiding truth, love, peace, and beauty and to give thanks. For it is not only about the destination back to love—the one unifying tune that holds, in complete union with it, all colors of the rainbow, all notes of the octave, all melodies, rich harmonics, lush overtones, including the discordant notes of our individual petty and ugly realities—but also just as much about wholeheartedly embracing and totally relishing this life, this awesome frequency-entangled and intensely personal odyssey of love through this absolutely beautiful musical mathematical universe.

Epilogue: Keeping It Simple

To date, an understanding of the laws that govern nature and the pleasure derived from the insights gained thus far have been the privilege of a select few: physicists, mathematicians, astronomers, cosmologists, biologists, and other scientists and intellectuals in their lofty circles. Since the inception of the quantum revolution, so much information has been acquired that the sheer enormity of it hinders lucid comprehension by the average person.

Quantum theory is without doubt the most successful theory ever to be studied and applied in theoretical and experimental physics and related sciences. The theory's utility in commerce and industry has changed the canvas of people's lives. Its rapid growth has aroused serious questions, provocative thoughts and lost wonderings that have confused a few members of the scientific community and have left common people like my friends and me in the lurch, even as we are thoroughly and delightedly immersed in the technology it has spawned.

Thankfully, much like the twine that recreational divers can use to trace back to their dive boats, the principle of vibration offers a steadfast beacon that will keep an awestruck novice from straying too far away from truth and clarity. My gut tells me that vibration, and its natural progression to frequency harmonics and resonance, is the unifying thread that runs through all that we know and those things we still wonder about. I have shed all fears of rejection and derision and I have followed my heart in my search for meaning and correlation between scientific discoveries and everybody's shared experiences of reality. Allow me to present my thoughts.

Everything vibrates. Vibrations are found in everything from the simplest element (hydrogen) to the whole universe. Vibration is central to the body of knowledge kept by the ancients and by the practitioners of occult sacred science. According to these teachings and traditions, all things in the universe originate from frequencies of sound (light) and harmonic principles: the first sound, *Om*, of Eastern practices; the light of *Eyn Sof* materializing into form through sound and number in Hebraic esoteric thought; and *logos,* the *Word* of the Christian Bible that was there in the beginning.

Everything that vibrates exists as energetic potentialities in clouds of probability waves arranged in harmonic order. The specific vibrational frequencies of probability waves dictate the interaction of atoms with one another to weave a harmonic interference pattern that the conscious mind holographically and geometrically discerns as a red rosebud, a beloved face, a haunting song, or as the harmonic pattern of the CMB. Thus, from ancient wisdom and modern discoveries, light in the zero-point field is information that human consciousness resonantly constructs as reality.

Everything that vibrates has intrinsic resonant frequencies. Resonance between frequencies of different entities is how one photon excites one electron in one chlorophyll molecule during plant photosynthesis, how the fine-tuned division of chromosomal DNA is accomplished, and how the nautilus, the sunflower, and the peacock get their complex shapes and fancy colorful patterns.

Resonance is the means by which plants and animals recognize kin from enemies. It is how bodily organs are imaged in medical diagnostics, how we communicate wirelessly, and how Tesla's pocket resonator made a Manhattan skyscraper vibrate. Resonance is how atomic, mechanical, and biological clocks synchronize, how the left and the right brain get entrained, and how minds meet. It is the principle behind rousing speech, moving poetry, touching music, passionate singing, brilliant acting, and expressive dancing.

Everything operates through resonant frequencies. Resonance and harmonics define the specific characteristics of molecules and tissues: their shapes, dimensions, densities, colors, scents, spatial separation, motion, function, and their macroscopic appearance on this earth as you and me with all our quirky and delightful attributes, including our ugly birthmarks, soulful eyes, and gorgeous dimples.

A person's resonance with electromagnetic information in the zero-point field orchestrates the materialization of potentialities upon the collapse of the probability wave function into all of our discerned reality in the brain microtubules. This is how the observer effect unfolds, focused attention gets results, prayers are answered, kindness begets kindness, synchronicities occur, savants and child prodigies exhibit inexplicable talent, seers perceive beyond space and time, and transcendent meditators reach a knowing past ordinary earthly understanding.

Seeking after and achieving resonance with zero-point energy potentialities is how we exercise free will and actively take part in directing the realization of our personal worlds—and thus be responsible for the course our own lives take.

Everything is entangled with everything else for all of perceived time and space. Any two or more particles, from photons to diamonds to human beings, that have ever been together show connectedness even after they have been separated across time and space. People have experienced quantum interconnectedness in one way or another in their daily lives, most commonly noticed in twins, family members, couples, close friends, and people and their pet animals and plants, as well as in the strong connection between patients and their physicians, counselors, and hypnotists. The encounter between the observer and the observed object involves a mutual resonant entanglement in an intensified oneness. We are one with one another, with the whole universe, and with the all-encompassing energy, whether or not we are aware of it; whether or not we accept and honor it. Through shared entwinement with pure consciousness, creatures on earth delight in one another's joys and feel one another's pain. This vibratory interconnectedness is undeniable and indissoluble all through eternity.

Harmonics and resonance order space, time, and gravity. Space is only a mathematical harmonic ordering of resonant frequencies in brain microtubules, interpreted by the mind as distance between the sun and the moon and between the comfortable couch and a ringing phone. There is no past, no present, and no future. The energies of objects and events we assign time to coexist as clouds of probabilities in the zero-point energy field, now and forever. Time is only a sequence of harmonic progression in the brain, given numbers such as seconds, hours, and years. Time starts and ends in the mind.

There is no need for strong nuclear forces to bind matter—from quarks to lovers to all of nature. The resonance of their composite frequencies holds them tightly together. The inhabitants of this planet and the heavens and the vibrations that compose them are precisely orchestrated mathematically and harmonically. The more resonant the frequencies of objects are to one another, the greater their mutual attraction.

I assert that resonance between objects is what we call gravity. My resonance with the earth keeps my feet on the ground. Nonetheless, when I go swimming, I do not sink in order to maintain contact with the earth. I resonate better with the water than with the earth, since my body is 99 percent water molecules. I also have air in my lungs that resonates better with the atmosphere than with the water and the earth. Therefore, I am able to stay afloat without having to expend energy. When I wish to swim underwater, I simply let my breath out, decreasing my resonance with the air and allowing my resonance with the water and with the earth to dominate. We can reassign the correlation of the term gravity from mass to frequency and resonance.

Mass as an expression of a concept is unnecessary. $E = mc^2$ is a space-time material discernment by the brain microtubules of the quantum-physical law $E = hv$. Someday, with enhanced and harmonically guided investigative techniques, we may be able to finally identify individual intrinsic resonant frequencies of objects in nature, if only just sufficiently to let go of the concept of a mass-dependent force of gravity. Whenever and whatever form that change comes about, it will come.

In the mathematics of frequency and resonance, one equation will describe nature. Following physical law, quantum mechanics and general relativity will resonantly meld into the unity of the singularity in the equations of vortex frequency, resonance, and entanglement. Perfect resonance and pure frequency must be what constitutes the unfathomable singularity of pure light: the standing wave core of the vortex of infinite energy of the zero-point field, the God of the believer.

Everything is energy, frequency, and vibration. All the above said, what remains at the fundamental level of reality is the wholeness of infinite energy of pure light. Pure frequency does not have to be "stringed" by string theory or "particulated" by quantum particle theory. The holographic whole is in everything, and everything is in the whole, including human consciousness. Pure light—the electromagnetic zero-point field energy—permeates, entangles with, and imparts distinguishing wave interference characteristics on everything there is in nature. Likewise, the holographic whole is informed back and gains an incalculable array of melodic variations from its vibrational entanglement with all things great and small, complex and simple. Thus, this source energy of pure light is lavishly embellished with the colorful rainbow of our experience in this awesome universe through an interactive instrument known as the human body.

In fairness, we are dealing with a duet of waves in the zero-point field of light frequencies and particles in the brain's orchestrated reduction of resonant electromagnetic waves. We must still use terms related to waves and particles until the advent of a richer language that can aptly describe life while fully incorporating both waves and particles. Alternatively, we might just go merrily on with the current descriptive duality, but with a tacit understanding that when we talk about matter, we are only describing mental images, not solid objects outside the mind.

Our reality is the material illusion of physical truth. Nature exists as a sea of waves of vibrational potentialities of being in the zero-point field that materialize upon electromagnetic encounter with human consciousness. The waves of probability that achieve the highest energy level upon entanglement with an observer's frequencies become manifest in the mind. On the whole, what we call reality—an elegantly set dining table, the sheer white curtains gently blowing in the breeze, the full moon rising up in the east, the distant sound of Fido barking at the moon, and Fido's master—is only a blend of harmonic crystallizations of frequencies electromagnetically presented to the microtubules, themselves also waves of probability. Through the amplifying effect of the energy of our own vibrations on zero-point probability waves, we directly influence the creation of our universe.

Consciousness is the fundamental element of the universe. Consciousness is an entity that is distinct from and yet entwined with the material organism. Consciousness is involved in all ways with our reality. This is a concept that is increasingly implied by scientific advances. But what is it? The answer from every corner is the same; only the language expressing it differs. Philosophers and sages proclaim and some scientists propose that consciousness is a fundamental physical non-material reality, an all-pervading and immeasurable sea of energy from which perceptions of matter in space and time are derived. Some quantum physicists speak of the zero-point energy field and stop because beyond there smacks of metaphysics, a tricky terrain which scientific angels fear to tread. Believers profess an omnipotent, omnipresent, and omniscient God of love, an all-encompassing energy from which everything emanates and to which everything must return, just as the sea is where every drop of rain comes from and to which it ultimately flows back.

Every attempt at linguistic expression of pure consciousness boils down to one and the same thing. It is energy. It is frequency. It is love. It is the utmost fundamental and absolute truth. And it offers a precious gift: we can tune in to its music in different ways and degrees. Some people are more adept at tuning into it, to a point of being able to reach a "knowing" of being one with it, having gone through many hours in a day and many years in a lifetime of meditative practice in close-to-ideal circumstances—a regimen many mortals cannot lay claim to.

Thankfully, the gift is generously available to everybody, the newcomers, the uninitiated, and the adepts alike. The tiniest yearning, the simplest devotion, is faithfully and rightly honored just the same. I am moved to recall the blessing told in Luke 23:42–43 (KJV) that the thief, hanging on a cross next to the crucified Christ, received when he sincerely implored, "Lord, remember me when thou cometh into thy kingdom." The God Incarnate answered him: "Verily I say unto thee, today shalt thou be with me in paradise." It was to this paradise of non-material perception, to this kingdom of the frequency of unconditional love, of light, that the repentant thief was invited to.

Everything in the universe is a unique eddy of resonant vibrations in the singularity of zero-point energy. In the center of the vortex, the frequencies are at their highest, most resonant, most stable, and still, united completely in the standing wave of the singularity. Less coherent frequencies may disconnect and then reconnect in accordance to harmonics, much like vortex knots in water or like the tune of a song that weaves in and out of the central theme or leaves the melody temporarily, only to rejoin once again by reason of resonance.

It follows that everything that consists of vortices of atoms is only a perceptible vortex of energy, including you, me, and the whole universe. All of our particular eddies of frequencies—our spirits, our one-of-a-kind symphonies—are part and parcel of the eternal energy singularity of the zero-point field and thus are themselves eternal. Our corruptible bodies—the structures that the mind interprets from the architecture of our uniquely individual music—whose assorted life experiences are observed unwaveringly by our spirits, are but fleeting illusions.

Love is the pure frequency of the singularity of the zero-point vortex of energy. Love is the frequency of that one sound, that one infinite light permeating everything. It is the only tune in the human being's unique vortex that endures through all the complex and ever-changing evolution of energy from the unchanging source. Love is the pure vibration that holds us all as one, including the many elaborate patterns of our harmonic entanglements—just as pure white light, when shone through a glass prism, breaks into all colors and hues of the rainbow, which, with another prism, come together to become pure white light again.

Love is the fundamental energy, the one frequency, the one pure light that transcendentally manifests in human consciousness. Much like a rich musical composition, the elaborate tapestry of pure light is intricately and inextricably woven wholly inclusive of all vibrant harmonics, subtle overtones, diverse interpretations, arrangements, and even dissonance into a divine symphony solely and completely responsible for begetting you and me and this beautiful musical mathematical universe.

You are loved.

Bibliography

In addition to the works listed below, several passages from the Holy Bible have been cited in this book, drawn from the King James Version (KJV) and the New King James Version (NKJV). These passages include Luke 23:42–43 and 24:36–39 (NKJV); Matthew 21:22, 22:36–40, and 25:40 (KJV); John 1:1–3 and 4:7–8 (KJV); Genesis 1:3 (KJV); and Acts 17:28 (KJV).

Alexander, Stephon. *The Jazz of Physics*. New York: Basic Books (2016).

American Institute of Physics Center for History of Physics. "Cosmic Journey: A History of Scientific Cosmology." Home page (n.d.). https://history.aip.org/exhibits/cosmology/index.htm.

American Physical Society. "This Month in Physics History: Einstein Predicts Stimulated Emission." APS News (August/September 2005). https://www.aps.org/publications/apsnews/200508/history.cfm.

Ardhuin, F., L. Gualtieri Land, and E. Stutzman. "How Ocean Waves Rock the Earth: Two Mechanisms Explain Microseisms with Periods 3 to 300 s." *Geophysical Research Letters*, 42 (2015): 765–72. https://agupubs.onlinelibrary.wiley.com/doi/10.1002/2014GL062782.

Arnold, James. "The Twin Paradox Resolved." *European Scientific Journal*, vol. 12, no. 3 (January 2016).

Aspect, Alain, Jean Dalibard, and Gerard Roger. "Experimental Test of Bell's Inequalities Using Time-Varying Analyzers." *Physical Review Letters*, vol. 49, no. 25 (1982): 1804–7.

Atrio-Barandela, F., et al. "A Measurement of Large-Scale Peculiar Velocities of Clusters of Galaxies: Results and Cosmological Implications." *Astrophysical Journal Letters*, vol. 675 (March 10, 2008): L57–L60. https://iopscience.iop.org/article/10.1086/592947. Aurilia, A., and E. Spallucci. "Fundamentals of Planckian Physics." Preprint manuscript (May

2013). https://www.cpp.edu/~aaurilia/Doc/research/planckian-physics.pdf.

Ayotte, Patrick, et al. "Infrared Spectroscopy of Negatively Charged Water Clusters: Evidence for a Linear Network." *Journal of Chemical Physics*, vol. 110, no. 13 (1999): 6268–77.

Backster, Grover Cleveland. *Primary Perception: Biocommunication with Plants, Living Foods, and Human Cells.* Anza, CA: White Rose Millennium Press (2003).

Badawy, A. *Ancient Egyptian architectural design. A study of the harmonic system.* CA: Univ. of California Press (1965).

Bandyopadhyay, Anirban. "Study of Opto-Electronic Properties of a Single Microtubule in the Microwave Regime." Final report for the Asian Office of Aerospace Research and Development (AOARD), grant FA2386-10-4059. National Institute for Materials Science, Tsukuba, Japan (March 5, 2014).

Berendt, Joachim-Ernst. *The World Is Sound: Nada Brahma: Music and the Landscape of Consciousness.* Rochester, VT: Destiny Books (1991).

Berger, Abi. "Positron Emission Tomography." *British Medical Journal*, vol. 326, no. 7404 (2003): 1449. https://doi.org/10.1136/bmj.326.7404.1449.

Bhagavad-Gita. Translated by Swami Prabhavananda and Christopher Isherwood. New York: Fall River Press (2010).

Bokulich, Alisa. "Bohr's Correspondence Principle." *Stanford Encyclopedia of Philosophy*, edited by Edward N. Zalta (Spring 2014 edition). https://plato.stanford.edu/archives/spr2014/entries/bohr-correspondence/.

Boyk, James. "There's Life above 20 Kilohertz! A Survey of Musical Instrument Spectra to 102.4 KHz." California Institute of Technology Music Lab, Pasadena. Unpublished manuscript (1992/1995). https://www.cco.caltech.edu/~boyk/spectra/spectra.htm.

Bryn, Brandon. "Science: The Breakthroughs of 2010 and Insights of the Decade." American Association for the Advancement of Science (December 16, 2010).

Buehner, Stephen Harrod. *The Lost Language of Plants.* White River Junction, VT: Chelsea Green (2002).

Buks, Eyal, et al. "Dephasing in Electron Interference by a 'Which-Path' Detector." *Nature*, vol. 391 (February 1998): 871–74. https://www.weizmann.ac.il/condmat/heiblum/papers/391871a0.pdf.

Capra, Fritjof. *The Tao of Physics: An Exploration of the Parallels between Modern Physics and Eastern Mysticism.* Boulder, CO: Shambhala (2000).

Chaisson, Eric J. *Cosmic Evolution: The Rise of Complexity in Nature.* Cambridge, MA: Harvard University Press (2002).

Chatrchyan, S. V., et al. "Observation of a New Boson at a Mass of 125 GeV with the CMS Experiment at the LHC." *Physics Letters B*, vol. 716, no. 1 (2012): 30–61. https://doi.org/10.1016/j.physletb.2012.08.021.

Cheney, Margaret. *Tesla: Man Out of Time*. New York: Touchstone (2001).

Chevalier, Gaétan. Review article: "Earthing: Health Implications of Reconnecting the Human Body to the Earth's Surface Electrons." *Journal of Environmental and Public Health*, vol. 2012 (2012). PMCID: PMC326507. http://dx.doi.org/10.1155/2012/291541.

Chopra, Deepak. "Reality and Consciousness: A View from the East." *Physics of Life Reviews*, vol. 11, no. 1 (March 2014): 81–2.

Chopra, Deepak, and Stuart Hameroff. "Consciousness—A Conversation with Deepak Chopra and Stuart Hameroff." *The Chopra Well* (July 15, 2013). YouTube.

Chou Kuo-Chen. "Low-Frequency Vibrations of DNA Molecules." *Biochemical Journal*, vol. 221, no. 1 (1984): 27–31.

Citizens Utility Board. "CUB's Guide to Energy Efficient Lighting." Home page (n.d.). http://citizensutilityboard.org.

Correa, Paulo N., and Alexandra N. Correa. "The Electron Mass-Energy Flux as a Deformable Nanometric Torus-Fine and Hyperfine Structures and Flux Topogeometries." *Aetherometric Theory of Synchronicity*, vol. 3, no. 4 (2011): 1–86.

Cowen, Robert C. "It's Official: Light Travels Exactly 299,762,458 Meters per Second." *Christian Monitor* (November 1, 1983). https://www.csmonitor.com/csm/contentmap/articles/1983-11-1.

Cox, Catherine Ann. "The Circadian Clock and the Cell Cycle." PhD thesis, University College London (2012). http://discovery.ucl.ac.uk/1348317/1/1348317.pdf.

Cramer, John G. "Sound of the Big Bang." University of Washington, Seattle. Unpublished manuscript (2013). https://faculty.washington.edu/jcramer/BBSound.html.

Cropper, William H. *The Quantum Physicists: And an Introduction to Their Physics*. New York: Oxford University Press (1970).

Davis, Abe, et al. "The Visual Microphone: Passive Recovery of Sound from Video." SIGGRAPH (2014). http://people.csail.mit.edu/mrub/VisualMic/.

Del Giudice, Emilio, Riccardo M. Pulselli, and Enzo Tiezzi. "Irreversible Processes and Quantum Field Theory: An Interplay for the Understanding of Ecosystem Dynamics." *Ecological Modelling*, vol. 220 (2009): 1874–9.

edge.org. "The Energy of Empty Space That Isn't Zero: A Talk with Lawrence M. Krauss" (2006). https://www.edge.org/conversation/lawrence_m_krauss-the-energy-of-empty-space-that-isnt-zero.

Eisaman, M. D., et al. "Shaping Quantum Pulses of Light via Coherent Atomic Memory." *Physical Review Letters*, vol. 93, no. 233602 (November 2004). https://doi.org/10.1103/PhysRevLett.93.233602.

Ellenberger, Henri F. *The Discovery of the Unconscious: The History and Evolution of Dynamic Psychiatry*. New York: Basic Books (1970).

Elliott, William J. "Circadian Variation in the Timing of Stroke Onset: A Meta-Analysis." *Stroke*, vol. 29 (1998): 992–6.

Encyclopedia.com. "Mesmer, Franz" (2008). http://www.encyclopedia.com/people/medicine/medicine-biographies/friedrich-anton-mesmer.

Everett, Hugh III, "The Theory of the Universal Wave Function." Princeton University (1957).

Falk, Dean, Frederick E. Lepore, and Adrienne Noe. "Cerebral Cortex of Einstein: A Description and Preliminary Analysis of Unpublished Photographs." *Brain*, vol. 136, no. 4 (2012): 1304–27.

Feynman, Richard. "Electricity in the Atmosphere." *The Feynman Lectures on Physics*, vol. II, chapter 9. Home page (n.d.). http://www.feynmanlectures.caltech.edu/.

Fischer, B. M., M. Walther, and J. P. Uhd. "Far-Infrared Vibrational Modes of DNA Components Studied by Terahertz Time-Domain Spectroscopy." Albert-Ludwig University, Freiburg, Germany (2002).

Franklin Institute. "Benjamin Franklin and the Kite Experiment." (n.d.). https://www.fi.edu/benjamin-franklin/kite-key-experiment.

Freedman, Stuart J., and John F. Clauser. "Experimental Test of Local Hidden-Variable Theories." *Physical Review Letters*, vol. 28, no. 14 (1972): 938–41.

Fuller, Franklin D., et al. "Vibronic Coherence in Oxygenic Photosynthesis." *Nature Chemistry*, vol. 6, no. 8 (2014): 706–11. https://www.ncbi.nlm.nih.gov/pubmed/25054941.

Gagliano, Monica, and Michael Renton. "Love Thy Neighbour: Facilitation through an Alternative Signalling Modality in Plants." *BMC Ecology*, vol. 13, no. 19 (2013).

Gagliano, Monica, et al. "Acoustic and Magnetic Communication in Plants: Is It Possible?" *Plant Signal Behavior*, vol. 7, no. 10 (October 1, 2012): 1346–8. https://www.tandfonline.com/doi/abs/10.4161/psb.27793.

————. "Out of Sight but Not out of Mind: Alternative Means of Communication in Plants." *PLoS ONE*, vol. 5 (2012). https://journals.plos.org/plosone/article?id=10.1371/journal.pone.0037382.

Galileo. *Dialogue concerning the Two Chief World Systems*. Translated by Stillman Drake. Oakland: University of California Press (1967).

Ghosh, Pallab. "God Particle Signal Is Simulated as Sound." *BBC News* (June 23, 2010).

Ghosh, Subrata, Satyajit Sahu, and Anirban Bandyopadhyay. "Evidence of Massive Global Synchronization and the Consciousness: Comment on 'Consciousness in the Universe: A Review of the "Orch OR" Theory' by Hameroff and Penrose." *Physics of Life Reviews*, vol. 11, no. 1 (2014).

Gibbs, Philip, and Don Koks. "What Is the Casimir Effect?" University of California, Riverside. Unpublished manuscript (2002). http://math.ucr.edu/home/baez/physics/Quantum/casimir.html.

Ginzburg, Vladimir B. "Double Helical and Double Toroidal Spiral Fields." *Speculations in Science and Technology*, vol. 21, no. 2 (1998): 79–89.

Goldstein, Sheldon. "Bohmian Mechanics." *Stanford Encyclopedia of Philosophy*, edited by Edward N. Zalta (2017). https://plato.stanford.edu/archives/sum2017/entries/qm-bohm.

Gomez Alvarez-Arenas, Tomas, et al. "Noncontact and Noninvasive Study of Plant Leaves Using Air-Coupled Ultrasounds." *Applied Physics Letters*, vol. 95, no. 19 (2009). http://dx.doi.org/10.1063/1.3263138.

Graham, Trent M., et al. "Superdense Teleportation Using Hyperentangled Photons." *Nature Communications*, vol. 6, no. 7185 (May 2015). https://www.nature.com/articles/ncomms8185.

Granek, Galena. "Poincaré's Light Signaling and Clock Synchronization Thought Experiment and Its Possible Inspiration to Einstein." University of Haifa (Israel) & American Institute of Physics (2006).

Greene, Brian. *The Elegant Universe: Superstrings, Hidden Dimensions, and the Quest for the Ultimate Theory*. New York: W. W. Norton & Company (2003).

————. *The Fabric of the Cosmos: Space, Time and the Texture of Reality*. New York: Vintage Books (2004).

————. *The Hidden Reality: Parallel Universes and the Deep Laws of the Cosmos*. New York: Vintage Books (2011).

Griffin, David Ray. *Consciousness as a Subjective Form: Whitehead's Nonreductionist Naturalism*. AnthonyFlood.com. Unpublished manuscript (November 12, 2009). www.anthonyflood.com/griffinconsciousness.htm.

Hackh, W. D. "The Classification of Chemical Elements." *Scientific American* (March 8, 1919). https://www.scientificamerican.com/article/the-classification-of-the-chemical/.

Hafele, Joseph C., and Richard E. Keating. "Around-the-World Atomic Clocks: Predicted Relativistic Time Gains." *Science*, vol. 177, no. 4044 (1972): 166–68. https://science.sciencemag.org/content/177/4044/166.

Hameroff, Stuart. Home page (2014). www.quantumconsciousness.org.

Hameroff, Stuart, and Roger Penrose. "Consciousness in the Universe: A Review of the 'Orch OR' Theory." *Physics of Life Reviews*, vol. 11, no. 1 (March 2014): 39–78. https://www.sciencedirect.com/science/article/pii/S1571064513001188.

Hartle, James B. "Theories of Everything and Hawking's Wave Function of the Universe." University of California, Santa Barbara. Preprint manuscript (n.d.). https://cds.cern.ch/record/581310/files/0209047.pdf.

Hawking, Stephen, and Leonard Mlodinow. *A Briefer History of Time*. New York: Bantam Dell (2008).

Hawkins, David R. *Power vs. Force*. West Sedona, AZ: Veritas Publishing (1998).

Herivel, John. "Christiaan Huygens—Dutch Scientist and Mathematician." *Encyclopedia Britannica*. https://www.britannica.com/biography/Christiaan-Huygens.

Hu, Wayne, and Martin White. "The Cosmic Symphony." Special report, *Scientific American* (February 2004): 44–53. http://background.uchicago.edu/~whu/Papers/HuWhi04.pdf.

Hubble, Edwin. "A Relation between Distance and Radial Velocity among Extra-Galactic Nebulae." *Proceedings of the National Academy of Sciences*, vol. 15 (1929): 168–73.

HubbleSite. "Black Hole–Powered Jet of Electrons and Sub-Atomic Particles Streams from Center of Galaxy M87" (July 6, 2000). http://hubblesite.org/image/968/gallery/50-exotic.

Illich, Ivan. "Hospitality and Pain." Presented to the McCormick Theological Seminary (1987).

Isaacson, Walter. *Einstein: His Life and Universe*. New York: Simon & Schuster (2008).

Jabr, Ferris. "How Does a Venus Flytrap Work?" NYU Scienceline (2010). http://scienceline.org/2010/03/how-does-a-venus-flytrap-work.

Jacquot, Jeremy. "Numbers: Nuclear Weapons, from Making a Bomb to Making a Stockpile to Making Peace." *Discover Magazine* (October 23, 2010).

Jeans, James. *The Mysterious Universe*. New York: Kessinger Legacy Reprints (1930; reissued 2009).

Julien, Claude. "The Enigma of Mayer Waves: Facts and Models." *Cardiovascular Research*, vol. 70, no. 1 (2006): 12–21. https://doi.org/10.1016/j.cardiores.2005.11.008.

Kaku, Michio. *Parallel Worlds: A Journey through Creation, Higher Dimensions, and the Future of the Cosmos.* New York: Anchor Books (2006).

Kalyani, Bangalore G., et al. "Neurohemodynamic Correlates of 'OM' Chanting: A Pilot Functional Magnetic Resonance Imaging Study." *International Journal of Yoga*, vol. 4, no. 1 (2011): 3–6.

Kavli Institute for the Physics and Mathematics of the Universe (Kavli IPMU) Subaru Telescope. "Hyper Suprime-Cam Ushers in a New Era of Observational Astronomy." http://www.ipmu.jp/node/1376.

Ketterle, Wolfgang. Nobel lecture. "When Atoms Behave as Waves: Bose-Einstein Condensation and the Atom Laser." *Review of Modern Physics*, vol. 74, no. 1131 (November 20, 2002). https://journals.aps.org/rmp/abstract/10.1103/RevModPhys.74.1131.

Kleckner, Dustin, and William T. M. Irvine. "Creation and Dynamics of Knotted Vortices." *Nature Physics*, vol. 9 (2013): 253–8. https://www.nature.com/articles/nphys2560.

Krotz, Dan. "The Secrets of a Cell's Skeleton at 8-Angstrom Resolution." Berkeley Lab Science Beat (December 17, 2002). https://www2.lbl.gov/Science-Articles/Archive/LSD-8-Angstrom-res.html.

Kruglinksi, Susan, and Oliver Chanarin. "Roger Penrose Says Physics Is Wrong, from String Theory to Quantum Mechanics." *Discover Magazine* interview (September 2009).

Kuþera, Ondrej, and Michal Cifra. "Cell-to-Cell Signaling through Light: Just a Ghost of a Chance?" *Cell Communication and Signaling*, vol. 11, no. 87 (2013). https://biosignaling.biomedcentral.com/articles/10.1186/1478-811X-11-87.

Kumar, Manjit. *Quantum: Einstein, Bohr, and the Great Debate about the Nature of Reality.* New York: W. W. Norton & Company (2008).

Kumar, Sanjay, et al. "Meditation on OM: Relevance from Ancient Texts and Contemporary Science." *International Journal of Yoga*, vol. 3, no. 1 (2010): 2–5.

Lanza, Robert, and Bob Berman. *Biocentrism: How Life and Consciousness Are the Keys to Understanding the Universe.* Dallas: Benbella Books (2009).

Lasky, Ronald C. "How Does Relativity Theory Resolve the Twin Paradox?" *Scientific American* (September 2002).

Lauterwasser, Alexander. *Water Sound Images: The Creative Music of the Universe.* Eliot, ME: Macromedia Publishing (2011).

Lee, K. C., et al. "Entangling Macroscopic Diamonds at Room Temperature." *Science*, vol. 334, no. 6060 (2011): 12531256. https://science.sciencemag.org/content/334/6060/1253.abstract?sid=3e228e8d-5dbc-4279-99ce-7a4a15fa51e5.

Leet, Leonora. *The Secret Doctrine of the Kabbalah*. Rochester, VT: Inner Traditions International (1999).

Lenger, Karin. "Homeopathic Potencies Identified by a New Magnetic Resonance Method: Homeopathy, an Energetic Medicine." *Subtle Energies & Energy Medicine*, vol. 15 (2005).

Levich, A. P. "A Substantial Interpretation of [the] N. A. Kozyrev Conception of Time." *World Scientific* (1996): 1–42. https://www.scribd.com/doc/133048207/A-Substantial-Interpretation-of-N-A-Kozyrev-Conception-of-Time-A-P-Levich#.

Li Y, Che Z, Quan W, et al. Diagnostic outcomes of magnetocardiography in patients with coronary artery disease. *Int J Clin Exp Med*. 2015;8(2):2441–2446. Published 2015 Feb 15.

Lincoln, Don. "The Good Vibrations of Quantum Field Theories." *NOVA* (August 5, 2013). http://www.pbs.org/wgbh/nova/blogs/physics/2013/08/the-good-vibrations-of-quantum-field-theories/.

Liu Zhihua and Chu Guiyan. "Chronobiology in Mammalian Health." *Molecular Biology Reports*, vol. 40, no. 3 (2013): 2491–501.

Lodish, H., et al. *Molecular Cell Biology*. 4th edition, section 19.1: "Microtubule Structures." New York: W. H. Freeman (2000). https://www.ncbi.nlm.nih.gov/books/NBK21580/.

Lumen. "Huygens' Principle." Boundless Physics (n.d.). https://courses.lumenlearning.com/boundless-physics/chapter/diffraction/.

Luminet, Jean-Pierre. "Editorial Note to 'A Homogeneous Universe of Constant Mass and Increasing Radius Accounting for the Radial Velocity of Extra-Galactic Nebula' by Georges Lemaître (1927)." Paris: Paris Diderot University (2013).

Madl, P., et al. "Evidence of Coherent Dynamics in Water Droplets of Waterfalls." *Water*, vol. 5 (July 20, 2013). http://www.waterjournal.org/volume-5/madl-summary.

Map of Life. "Vibrational Communication in Animals" (2009). http://www.mapoflife.org/topics/topic_458_vibrational-communication-in-animals/.

Martin, Thomas Commerford. *The Inventions, Researches and Writings of Nikola Tesla*. New York (1893; reprinted and published by Barnes & Noble, 1995).

Maury, Eleonore, et al. "Circadian Rhythms and Metabolic Syndrome: From Experimental Genetics to Human Disease." *Circulation Research*, vol. 106, no. 3 (2010): 447–62.

Mazzoccoli, Gianluigi, et al. "Differential Patterns in the Periodicity and Dynamics of Clock Gene Expression in Mouse Liver and Stomach." *Journal of Biological and Medical Rhythm Research*, vol. 29, no. 10 (2012). http://dx.doi.org/10.3109/07420528.2012.728662.

McCraty, Rollin, et al. "The Electricity of Touch: Detection and Measurement of Cardiac Energy Exchange between People." In *Brain and Values: Is a Biological Science of Values Possible?*, edited by Karl Pribram. Mahwah, NJ: Lawrence Erlbaum (1998): 359–79.

McCraty, Rollin, Annette Deyhle, and Doc Childre. "Global Coherence Initiative: Creating a Coherent Planetary Standing Wave." *Global Advances in Health and Medicine*, vol. 1, no. 1 (March 2012): 64–77. https://journals.sagepub.com/doi/10.7453/gahmj.2012.1.1.013.

McCraty, Rollin, and Fred Shaffer. "Hear Rate Variability: New Perspectives on Physiological Mechanisms, Assessment of Self-Regulatory Capacity, and Health Risk." *Global Advancements in Health and Medicine*, vol. 4, no. 1 (2015): 46–61.

McGee, Ryan, et al. "Sonifying the Cosmic Microwave Background." University of California, Santa Barbara: *17th International Conference on Auditory Display (ICAD-2011)*, Budapest, Hungary (June 20–24, 2011).

Megidish, Eli, et al. "Entanglement between Photons That Have Never Coexisted." *Physics Review Letters*, vol. 110, no. 210403 (May 2013). https://doi.org/10.1103/PhysRevLett.110.210403.

Meumandi, Assad. "Music, Medicine, Healing, and the Genome Project." *Psychiatry*, vol. 6, no. 9 (2009): 43–5.

Meyl, Konstantin. "About Vortex Physics and Vortex Losses." *Journal of Vortex Science and Technology*, vol. 1 (2012).

———. "DNA and Cell Resonance: Magnetic Waves Enable Cell Communication." *DNA Cell Biology*, vol. 4 (April 31, 2012): 422–6.

Milonni, P. W. "Why Spontaneous Emission?" *American Journal of Physics*, vol. 52, no. 340 (1984). http://dx.doi.org/10.1119/1.13886.

Moravec, C. S., and M. G. McKee. "Biofeedback in the Treatment of Heart Disease." *Cleveland Clinic Journal of Medicine*, vol. 78, suppl. 1 (2011): S20–3.

Moser, Edvard, May-Britt Moser, and Yassir Roudi. "Network Mechanisms of Grid Cells." *Philosophical Transactions of the Royal Society B*, vol. 369, no. 20120511 (2014). https://royalsocietypublishing.org/doi/full/10.1098/rstb.2012.0511.

Muehsam, David, and Carlo Ventura. "Life Rhythm as a Symphony of Oscillatory Patterns: Electromagnetic Energy and Sound Vibration Modulates Gene

Expression for Biological Signaling and Healing." *Global Advances in Health and Medicine*, vol. 3, no. 2 (2014): 40–55. https://journals.sagepub.com/doi/10.7453/gahmj.2014.008.

Murphy, Guillermo P., and Susan Dudley. "Kin Recognition: Competition and Cooperation in Impatiens (Balsaminaceae)." *American Journal of Botany*, vol. 96, no. 11 (1990).

NASA Chandra. "Cassiopeia A Comes Alive across Time and Space" (2009). www.nasa.gov/mission_pages/chandra/news/09-001.html.

NASA Goddard Space Flight Center. "Edwin P. Hubble" (n.d.). https://asd.gsfc.nasa.gov/archive/hubble/overview/hubble_bio.html.

———. "Universe Is Expanding" (1929). https://imagine.gsfc.nasa.gov/educators/programs/cosmictimes/online_editio n/1929/index.html.

NASA Hubble. "Hubble Reveals Observable Universe Contains 10 Times More Galaxies Than Previously Thought." *Cosmic Times* (October 13, 2016). https://www.nasa.gov/feature/goddard/2016/hubble-reveals-observable-universe-contains-10-times-more-galaxies-than-previ-ously-thought.

———. "Hubble Team Breaks Cosmic Distance Record" (2016). https://www.nasa.gov/feature/goddard/2016/hubble-team-breaks-cosmic-distance-record.

NASA Jet Propulsion Laboratory, California Institute of Technology. "Missions / Cassini-Huygens." Home page (n.d.). http://saturn.jpl.nasa.gov.

NASA Science. "Stars" (n.d.). https://science.nasa.gov/astrophysics/focus-areas/how-do-stars-form-and-evolve/.

National Cancer Institute (NCI). "Electromagnetic Fields and Cancer" (n.d.). https://www.cancer.gov/about-cancer/causes-prevention/risk/radiation/electromagnetic-fields-fact-sheet.

———. "Cell Phones and Cancer Risk Fact Sheet" (n.d.). https://www.cancer.gov/about-cancer/causes-prevention/risk/radiation/cell-phones-fact-sheet.

———. "Computed Tomography (CT) Scans and Cancer Fact Sheet" (n.d.). https://www.cancer.gov/about-cancer/diagnosis-staging/ct-scans-fact-sheet.

National Geographic Society. "Supernovae" (n.d.). http://www.nationalgeographic.com/science/space/universe/supernovae/.

O'Connor, John Joseph, and E. F. Robertson. "Galileo Galilei." University of Saint Andrews (2002). http://www-history.mcs.st-andrews.ac.uk/Biographies/Galileo.html.

Oliveira, Arnaldo. "Electroacupuncture According to Voll: Historical Background and Literature Review." *Journal of Acupuncture and Oriental Medicine* (January 2016).

Oppenheimer, Stephen. "The Broken Heart: Noninvasive Measurement of Cardiac Autonomic Tone." *Postgraduate Medical Journal*, vol. 68, no. 806 (December 1992): 939–41.

O'Reilly, Edward J., and Alexandra Olaya-Castro. "Nonclassicality of the Molecular Vibrations Assisting Exciton Energy Transfer at Room Temperature." *Nature Communications*, vol. 5, no. 3012 (2014).

"Overtones and Harmonics." HyperPhysics, Georgia State University (2016). http://hyperphysics.phy-astr.gsu.edu.

Øyvind, Grøn. "The Twin Paradox, and the Principle of Relativity." Unpublished manuscript. Oslo University College (n.d.). http://www.uio.no/studier/emner/matnat/fys/FYS4160/v13/the-twin-paradox-and-the-principle-of-relativity.pdf.

Paddison, Sara. *The Hidden Power of the Heart*. Boulder Creek, CO: Planetary Publications (1992).

Parker, Barry R. *The Vindication of the Big Bang: Breakthroughs and Barriers*. New York: Plenum Press (1993).

Peat, F. David, and John Briggs. "Interview with David Bohm." Originally published in *Omni* (1987). https://www.bibliotecapleyades.net/esp_paradigmaholo08.htm.

Perelman, M. E., and G. M. Rubinstein. "Ultrasound Vibrations of Plant Cell Membranes: Water Lift in Trees, Electrical Phenomena." Electronic preprint (n.d.). Jerusalem: Hebrew University of Jerusalem. https://arxiv.org/ftp/physics/papers/0611/0611133.pdf.

Piazza, L., et al. "Simultaneous Observation of the Quantization and the Interference Pattern of a Plasmonic Near-Field." *Nature Communications*, vol. 6, no. 6407 (2015). https://www.nature.com/articles/ncomms7407.

Pike, Albert. *Pythagoras & Hermes*. Lynwood, MA: Holmes Publishing Group (1999).

Pribram, Karl. "Holonomic Brain Theory." *Scholarpedia*, vol. 2, no. 5 (2007): 2735.

Prigogine, Ilya, and Isabelle Stengers. *Order Out of Chaos: Man's New Dialogue with Nature*. New York: Bantam Books (1984).

"Quantum Jitters Could Form the Basis of Evolution." Duke University (2015). https://phys.org/news/2015-03-quantum-jitters-basis-evolution-cancer.html.

Randall, J. M., R. T. Matthews, and M. A. Stiles. "Resonant Frequencies of Standing Humans." *Ergonomics*, vol. 40, no. 9 (1997).

Randall, Lisa. *Warped Passages: Unraveling the Mysteries of the Universe's Hidden Dimensions*. New York: Harper Perennial (2006).

Rein, Glen, and Rollin McCraty. "Modulation of DNA by Coherent Heart Frequencies." *Proceedings of the Third Annual Conference of the International Society for the Study of Subtle Energies & Energy Medicine, Monterey, CA* (1993).

———. "Structural Changes in Water and DNA Associated with New Physiologically Measured States." *Journal of Scientific Exploration*, vol. 8, no. 3 (1994): 438.

Reiter, G. F., et al. "Evidence of a New Quantum State of Nano-Confined Water." Preprint manuscript (January 27, 2011).

Rice, Stuart A., and Joshua Jortner. *James Franck 1882–1964: A Biographical Memoir*. Washington, DC: National Academy of Sciences (2010). http://jfi.uchicago.edu/resources/history/RiceJortner_2010.pdf.

Rieper, Elisabeth, Janet Anders, and Vlatko Vedral. "Quantum Entanglement between the Electron Clouds of Nucleic Acids in DNA." Preprint manuscript (2010). https://arxiv.org/pdf/1006.4053.pdf.

Ringermacher, Harry I., and Laurence R. Mead. "Is the Universe Ringing Like a Crystal Glass?" (2016). Preprint manuscript. http://www.ringermacher.com/wp-content/uploads/2016/08/OscillationPPR_AJ2_5-21-14_Ringermacher.pdf.

Romeo, Nelide, Olivier Gallo, and Giuseppe Tagarelli. "From Disease to Holiness: Religious-Based Health Remedies of Italian Folk Medicine (XIX–XX Century)." *Journal of Ethnobiology and Ethnomedicine*, vol. 11 (2015): 50.

Rosenblum, Bruce, and Fred Kuttner. *Quantum Enigma: Physics Encounters Consciousness*. New York: Oxford University Press (2011).

Rubin, Vera, and W. Kent Ford. "Rotation of the Andromeda Nebula from a Spectroscopic Survey of Emission Regions." *Astrophysical Journal*, vol. 159 (1970): 379.

Russell, Peter. From *Science to God: A Physicist's Journey into the Mystery of Consciousness*. Novato, CA: New World Library (2002).

Sagan, Carl. *Cosmos*. New York: Random House (1980).

Sahu, Satyajit, et al. "Atomic Water Channel Controlling Remarkable Properties of a Single Brain Microtubule: Correlating Single Protein to Its Supramolecular Assembly." *Biosensors and Bioelectronics*, vol. 15, no. 47 (2013):141–8. https://www.ncbi.nlm.nih.gov/pubmed/23567633.

Salaman, Clement, et al. (translators). *The Way of Hermes: New Translations of* The Corpus Hermeticum *and* The Definitions of Hermes Trismegistus to Asclepius. Rochester, VT: Inner Traditions International (2000).

Sanders, Robert. "Milky Way Galaxy Is Warped and Vibrating Like a Drum." UC Berkeley News Media Relations (2006).

———. "Most Precise Test Yet of Einstein's Gravitational Redshift." UC Berkeley Media Relations (2010).

Satoshi, Y., et al. "Wide Range Tuning of Resonant Frequency for a Vortex Core in a Regular Triangle Magnet." *Scientific Reports*, vol. 3 (2013).

Schibler, Ueli. "Circadian Time Keeping: The Daily Ups and Downs of Genes, Cells, and Organisms." Review. *Progress in Brain Research*, vol. 153 (2006): 271–82.

———. "The Daily Rhythms of Genes, Cells and Organs." University of Geneva (2005).

Senzaki, Nyogen, and Ruth McCandless. *The Iron Flute: 100 Zen Koans.* Kindle edition. Tuttle Publishing (2011).

Shealy, Norman, et al. "EEG Alterations during Absent Healing." *Subtle Energies & Energy Medicine*, vol. 11, no. 3 (2000): 241–48.

Shikhobalov, Lavrenty. "The Fundamentals of N. A. Kozyrev's Causal Mechanics." *On the Way to Understanding the Time Phenomenon: The Constructions of Time in Natural Science, part 2: The "Active" Properties of Time According to N. A. Kozyrev.* Singapore, New Jersey, London, and Hong Kong: World Scientific (1996): 43–59.

Sleight, Peter. "Cardiovascular Effects of Music by Entraining Cardiovascular Autonomic Rhythms Music Therapy Update: Tailored to Each Person, or Does One Size Fit All?" *Netherlands Heart Journal*, vol. 21, no. 2 (2013): 99–100.

Sleight, Peter, et al. "Physiology and Pathophysiology of Heart Rate and Blood Pressure Variability in Humans: Is Power Spectral Analysis Largely an Index of Baroreflex Gain?" *Clinical Science*, vol. 88 (1995): 103–9.

Smoot, George F. "Cosmic Microwave Background Radiation Anisotropies: Their Discovery and Utilization." Nobel lecture, December 8, 2006. Stockholm (2007).

Starwynn, Darren. "Vibrational Medicine for Acupuncturists, Part One: Light and Electricity." *Acupuncture Today*, vol. 4, no. 7 (2003).

Stodolna, Aneta, et al. "Hydrogen Atom under Magnification: Direct Observation of the Nodal Structure of Stark States." *Physical Review Letters*, vol. 110, no. 2013001 (2013). https://doi.org/10.1103/Physical ReviewLetters.110.213001.

Storoy, David. "David Bohm, Implicate Order and Holomovement." *Science and Nonduality* (2014). https://www.scienceandnonduality.com/david-bohm-implicate-order-and-holomovement/.

Straumann, Norbert. "Fritz Zwicky: An Extraordinary Astrophysicist." *Swiss Physical Society* (2013).

Subaru Telescope Facility, Hilo, Hawaii. "Cosmic Giants Shed New Light on Dark Matter." Originally published by the University of Tokyo. *Astronomy* (June 13, 2013). http://www.astronomy.com/news/2013/06/cosmic-giants-shed-new-light-on-dark-matter.

Talbot, Michael. *The Holographic Universe: The Revolutionary Theory of Reality That Explains the Latest Frontiers of Physics, the Paranormal Abilities of the Mind, the Unsolved Riddles of Brain and Body.* New York: Harper Perennial (1991).

Tebecis, A. K. "A Controlled Study of the EEG during Transcendental Meditation: Comparison with Hypnosis." *Folia psychiatrica et neurologica japonica*, vol. 29, no. 4 (1975): 305–13.

Thompson, Laird. "Vesto Slipher and the First Galaxy Redshifts." Preprint manuscript (2011). arXiv:1108.4864.

Tompkins, Peter, and Christopher Bird. *The Secret Life of Plants.* New York: Harper Collins Publisher (1973).

————. *Secrets of the Soil.* Eagle River, CA: Earthpulse Press (2002).

Tsodyks, M., Asher Uziel, and Henry Markram. "Synchrony Generation in Recurrent Networks with Frequency-Dependent Synapses." *Journal of Neuroscience*, vol. 20, no. 1 (2000).

Udo, Ryuta, et al. "The Role of Clock[s] in the Plasticity of Circadian Entrainment." *Biochemical and Biophysical Research Communications*, vol. 318, no. 4 (2004): 893–8.

University of Michigan. "Deep within Spinach Leaves, Vibrations Enhance Efficiency of Photosynthesis." *Science Daily* (2014). www.sciencedaily.com/releases/2014/07/140713155502.htm.

Villasante, A., et al. "Binding of Microtubule Protein to DNA and Chromatin: Possibility of Simultaneous Linkage of Microtubule to Nucleic Acid and Assembly of the Microtubule Structure." *Nucleic Acids Research*, vol. 9, no. 4 (1981): 895–908.

Vinciguerra, Manlio, Maria Florencia Tevey, and Gianluigi Mazzoccoli. "A Ticking Clock Links Metabolic Pathways and Organ Systems Function in Health and Disease." *Clinical and Experimental Medicine*, vol. 14, no. 2 (2014): 133–40.

Volkow, Nora, et al. "Effects of Cell Phone Radiofrequency Signal Exposure on Brain Glucose Metabolism." *Journal of the American Medical Association*, vol. 305, no. 8 (2011): 808–13. https://jamanetwork.com/journals/jama/fullarticle/645813.

Ward, Tarsha, et al. "Regulation of a Dynamic Interaction between Two Microtubule-Binding Proteins, EB1 and TIP150, by the Mitotic p300/CBP-

Associated Factor (PCAF) Orchestrates Kinetochore Microtubule Plasticity and Chromosome Stability during Mitosis." *Journal of Biological Chemistry*, vol. 288, no. 22 (2013): 15771–85.

Weinberg, Steven. *The First Three Minutes*. New York: Basic Books (1977).

Westinghouse. "History of George Westinghouse—Innovation Changing the World" (n.d.). http://www.westinghousenuclear.com/About/History.

Wilcock, David. *The Source Field Investigations*. New York: Penguin Group (2012).

Wolff, Milo. "Solving Nature's Mystery: On the Spherical Wave Structure of Matter and the Origin of Natural Laws. Explaining the Particle Wave Duality of Light and Matter with the Wave Structure of Matter (WSM)." Unpublished manuscript (2003). https://www.spaceandmotion.com/Wolff-Wave-Structure-Matter.htm.

World Health Organization. "Electromagnetic Fields and Public Health: Base Stations and Wireless Technologies." WHO (2006). www.who.int/peh-emf/publications/facts/fs304/en/.

Yutaka, Sumino, et al. "Large-Scale Vortex Lattice Emerging from Collectively Moving Microtubules." *Nature*, vol. 483 (2012): 448–52.

Zimecki, Michael. "The Lunar Cycle: Effects on Human and Animal Behavior and Physiology." Postępy Higieny i Medycyny Doświadczalnej (Advances in Hygiene and Experimental Medicine), vol. 60 (2006): 1

www.ingramcontent.com/pod-product-compliance
Lightning Source LLC
Chambersburg PA
CBHW071411180526
45170CB00001B/62